PRINCIPLES OF ARTIFICIAL NEURAL NETWORKS

2nd Edition

ADVANCED SERIES IN CIRCUITS AND SYSTEMS

Editor-in-Charge: **Wai-Kai Chen** (Univ. Illinois, Chicago, USA)
Associate Editor: **Dieter A. Mlynski** (Univ. Karlsruhe, Germany)

Published

Vol. 1: Interval Methods for Circuit Analysis
by *L. V. Kolev*

Vol. 2: Network Scattering Parameters
by *R. Mavaddat*

Vol. 3: Principles of Artificial Neural Networks
by *D Graupe*

Vol. 4: Computer-Aided Design of Communication Networks
by *Y-S Zhu & W K Chen*

Vol. 5: Feedback Networks: Theory & Circuit Applications
by *J Choma & W K Chen*

Advanced Series on Circuits and Systems – Vol. 6

PRINCIPLES OF ARTIFICIAL NEURAL NETWORKS

2nd Edition

Daniel Graupe

University of Illinois, Chicago, USA

NEW JERSEY • LONDON • SINGAPORE • BEIJING • SHANGHAI • HONG KONG • TAIPEI • CHENNAI

Published by

World Scientific Publishing Co. Pte. Ltd.
5 Toh Tuck Link, Singapore 596224
USA office: 27 Warren Street, Suite 401-402, Hackensack, NJ 07601
UK office: 57 Shelton Street, Covent Garden, London WC2H 9HE

British Library Cataloguing-in-Publication Data
A catalogue record for this book is available from the British Library.

PRINCIPLES OF ARTIFICIAL NEURAL NETWORKS (2nd Edition)
Advanced Series on Circuits and Systems – Vol. 6

ISBN-13 978-981-270-624-9
ISBN-10 981-270-624-0

Printed in Singapore by B & JO Enterprise

Dedicated to the memory of my parents,
to my wife Dalia,
to our children, our daughters-in-law and our grandchildren
It is also dedicated to the memory of Dr. Kate H Kohn

Acknowledgments

I am most thankful to Hubert Kordylewski of the Department of Electrical Engineering and Computer Science of the University of Illinois at Chicago for his help towards the development of LAMSTAR network of Chapter 13 of this text. I am grateful to several students who attended my classes on Neural Network at the Department of Electrical Engineering and Computer Science of the University of Illinois at Chicago over the past fourteen years and who allowed me to append programs they wrote as part of homework assignments and course projects to various chapters of this book. They are Vasanth Arunachalam, Sang Lee, Maxim Kolesnikov, Hubert Kordylewski, Maha Nujeimo, Michele Panzeri, Padmagandha Sahoo, Daniele Scarpazza, Sanjeeb Shah and Yunde Zhong.

I am deeply indebted to the memory of Dr. Kate H. Kohn of Michael Reese Hospital, Chicago and of the College of Medicine of the University of Illinois at Chicago and to Dr. Boris Vern of the College of Medicine of the University of Illinois at Chicago for reviewing parts of the manuscript of this text and for their helpful comments.

Ms. Barbara Aman and the production and editorial staff at World Scientific Publishing Company in Singapore were extremely helpful and patient with me during all phases of preparing this book for print.

Preface to the First Edition

This book evolved from the lecture notes of a first-year graduate course entitled "Neural Networks" which I taught at the Department of Electrical Engineering and Computer Science of the University of Illinois at Chicago over the years 1990–1996. Whereas that course was a first-year graduate course, several Senior-Year undergraduate students from different engineering departments, attended it with little difficulty. It was mainly for historical and scheduling reasons that the course was a graduate course, since no such course existed in our program of studies and in the curricula of most U.S. universities in the Senior Year Undergraduate program. I therefore consider this book, which closely follows these lecture notes, to be suitable for such undergraduate students. Furthermore, it should be applicable to students at that level from essentially every science and engineering University department. Its prerequisites are the mathematical fundamentals in terms of some linear algebra and calculus, and computational programming skills (not limited to a particular programming language) that all such students possess.

Indeed, I strongly believe that Neural Networks are a field of both intellectual interest and practical value to all such students and young professionals. Artificial neural networks not only provide an understanding into an important computational architecture and methodology, but they also provide an understanding (very simplified, of course) of the mechanism of the biological neural network.

Neural networks were until recently considered as a "toy" by many computer engineers and business executives. This was probably somewhat justified in the past, since neural nets could at best apply to small memories that were analyzable just as successfully by other computational tools. I believe (and I tried in the later chapters below to give some demonstration to support this belief) that neural networks are indeed a valid, and presently, the only efficient tool, to deal with very large memories.

The beauty of such nets is that they can allow and will in the near-future allow, for instance, a computer user to overcome slight errors in representation, in programming (missing a trivial but essential command such as a period or any other symbol or character) and yet have the computer execute the command. This will obviously require a neural network buffer between the keyboard and the main programs. It should allow browsing through the Internet with both fun and efficiency. Advances in VLSI realizations of neural networks should allow in the coming years many concrete applications in control, communications and medical devices, including in artificial limbs and organs and in neural prostheses, such as neuromuscular stimulation aids in certain paralysis situations.

For me as a teacher, it was remarkable to see how students with no background in signal processing or pattern recognition could easily, a few weeks (10–15 hours) into the course, solve speech recognition, character identification and parameter estimation problems as in the case studies included in the text. Such computational capabilities make it clear to me that the merit in the neural network tool is huge. In any other class, students might need to spend many more hours in performing such tasks and will spend so much more computing time. Note that my students used only PCs for these tasks (for simulating all the networks concerned). Since the building blocks of neural nets are so simple, this becomes possible. And this simplicity is the main feature of neural networks: A house fly does not, to the best of my knowledge, use advanced calculus to recognize a pattern (food, danger), nor does its CNS computer work in picosecond-cycle times. Researches into neural networks try, therefore, to find out why this is so. This leads and led to neural network theory and development, and is the guiding light to be followed in this exciting field.

Daniel Graupe
Chicago, IL
January 1997

Preface to the Second Edition

The Second Edition contains certain changes and additions to the First Edition. Apart from corrections of typos and insertion of minor additional details that I considered to be helpful to the reader, I decided to interchange the order of Chapters 4 and 5 and to rewrite Chapter 13 so as to make it easier to apply the LAMSTAR neural network to practical applications. I also moved the Case Study 6.D to become Case Study 4.A, since it is essentially a Perceptron solution.

I consider the Case Studies important to a reader who wishes to see a concrete application of the neural networks considered in the text, including a complete source code for that particular application with explanations on organizing that application. Therefore, I replaced some of the older Case Studies with new ones with more detail and using most current coding languages (MATLAB, Java, C++). To allow better comparison between the various neural network architectures regarding performance, robustness and programming effort, all Chapters dealing with major networks have a Case Study to solve the same problem, namely, character recognition. Consequently, the Case studies 5.A (previously, 4.A, since the order of these chapters is interchanged), 6.A (previously, 6.C), 7.A, 8.A, have all been replaced with new and more detailed Case Studies, all on character recognition in a 6×6 grid. Case Studies on the same problem have been added to Chapter 9, 12 and 13 as Case Studies 9.A, 12.A and 13.A (the old Case Studies 9.A and 13.A now became 9.B and 13.B). Also, a Case Study 7.B on applying the Hopfield Network to the well known Traveling Salesman Problem (TSP) was added to Chapter 7. Other Case Studies remained as in the First Edition.

I hope that these updates will add to the readers' ability to better understand what Neural Networks can do, how they are applied and what the differences are between the different major architectures. I feel that this and the case studies with their source codes and the respective code-design details will help to fill a gap in the literature available to a graduate student or to an advanced undergraduate Senior who is interested to study artificial neural networks or to apply them.

Above all, the text should enable the reader to grasp the very broad range of problems to which neural networks are applicable, especially those that defy analysis and/or are very complex, such as in medicine or finance. It (and its Case Studies) should also help the reader to understand that this is both doable and rather easily programmable and executable.

Daniel Graupe
Chicago, IL
September 2006

Contents

Chapter 1

Introduction and Role of Artificial Neural Networks

Artificial neural networks are, as their name indicates, computational networks which attempt to simulate, in a gross manner, the networks of nerve cell (neurons) of the biological (human or animal) central nervous system. This simulation is a gross cell-by-cell (neuron-by-neuron, element-by-element) simulation. It borrows from the neurophysiological knowledge of biological neurons and of networks of such biological neurons. It thus differs from conventional (digital or analog) computing machines that serve to replace, enhance or speed-up human brain computation without regard to organization of the computing elements and of their networking. Still, we emphasize that the simulation afforded by neural networks is very gross.

Why then should we view artificial neural networks (denoted below as neural networks or ANNs) as more than an exercise in simulation? We must ask this question especially since, computationally (at least), a conventional digital computer can do everything that an artificial neural network can do.

The answer lies in two aspects of major importance. The neural network, by its simulating a biological neural network, is in fact a novel computer architecture *and* a novel algorithmization architecture relative to conventional computers. *It allows using very simple computational operations* (additions, multiplication and fundamental logic elements) to solve complex, mathematically ill-defined problems, nonlinear problems or stochastic problems. A conventional algorithm will employ complex sets of equations, and will apply to only a given problem and exactly to it. *The ANN will be* (a) *computationally* and algorithmically *very simple* and (b) it will have a *self-organizing feature* to allow it to hold for a wide range of problems.

For example, if a house fly avoids an obstacle or if a mouse avoids a cat, it certainly solves no differential equations on trajectories, nor does it employ complex pattern recognition algorithms. Its brain is very simple, yet it employs a few basic neuronal cells that fundamentally obey the structure of such cells in advanced animals and in man. The artificial neural network's solution will also aim at such (most likely not the same) simplicity. Albert Einstein stated that a solution or a model must be as simple as possible to fit the problem at hand. Biological systems, in order to be as efficient and as versatile as they certainly are despite their inherent slowness (their basic computational step takes about a millisecond versus less than a nanosecond in today's electronic computers), can only do so by converging to the simplest algorithmic architecture that is possible. Whereas high level mathematics and logic can yield a broad general frame for solutions and can be reduced to specific but complicated algorithmization, the neural network's design aims at utmost simplicity and utmost self-organization. A very simple base algorithmic structure lies behind a neural network, but it is one which is highly adaptable to a broad range of problems. We note that at the present state of neural networks their range of adaptability is limited. However, their design is guided to achieve this simplicity and self-organization by its gross simulation of the biological network that is (must be) guided by the same principles.

Another aspect of ANNs that is different and advantageous to conventional computers, at least potentially, is in *its high parallelity* (element-wise parallelity). A conventional digital computer is a *sequential machine.* If one transistor (out of many millions) fails, then the whole machine comes to a halt. In the adult human central nervous system, neurons in the thousands die out each year, whereas brain function is totally unaffected, except when cells at very few key locations should die and this in very large numbers (e.g., major strokes). This insensitivity to damage of few cells is due to the high parallelity of biological neural networks, in contrast to the said sequential design of conventional digital computers (or analog computers, in case of damage to a single operational amplifier or disconnections of a resistor or wire). The same redundancy feature applies to ANNs. However, since presently most ANNs are still simulated on conventional digital computers, this aspect of insensitivity to component failure does not hold. Still, there is an increased availability of ANN hardware in terms of integrated circuits consisting of hundreds and even thousands of ANN neurons on a single chip does hold. [cf. Jabri *et al.*, 1996, Hammerstom, 1990, Haykin, 1994]. In that case, the latter feature of ANNs.

In summary, the excitement in ANNs should not be limited to its greater resemblance to the human brain. Even its degree of self-organizing capability can be built into conventional digital computers using complicated artificial intelligence algorithms. The main contribution of ANNs is that, in its gross imitation of the biological neural network, it allows for very low level programming to allow solving complex problems, especially those that are non-analytical and/or nonlinear and/or nonstationary and/or stochastic, *and* to do so in a self-organizing manner that applies to a wide range of problems with no re-programming or other interference in the program itself. The insensitivity to partial hardware failure is another great attraction, but only when dedicated ANN hardware is used.

It is becoming widely accepted that the advent of ANN will open *new understanding into how* to *simplify* programming and algorithm design for a given end and for a wide range of ends. It should bring attention to the simplest algorithm *without*, of course, *dethroning advanced mathematics* and logic, whose role will always be supreme in mathematical understanding and which will always provide a systematic basis for eventual reduction to specifics.

What is always amazing to many students and to myself is that after six weeks of class, first year engineering graduate students of widely varying backgrounds with no prior background in neural networks or in signal processing or pattern recognition, were able to solve, individually and unassisted, problems of speech recognition, of pattern recognition and character recognition, which could adapt in seconds or in minutes to changes (with a range) in pronunciation or in pattern. They would, by the end of the one-semester course, all be able to demonstrate these programs running and adapting to such changes, using PC simulations of their respective ANNs. My experience is that the study time and the background to achieve the same results by conventional methods by far exceeds that achieved with ANNs.

This, to me, demonstrates the degree of simplicity and generality afforded by ANN; and therefore the potential of ANNs.

Obviously, if one is to solve a set of differential equations, one would not use an ANN, just as one will not ask the mouse or the cat to solve it. But problems of recognition, filtering and control would be problems suited for ANNs. As always, no tool or discipline can be expected to do it all. And then, ANNs are certainly at their infancy. They started in the 1950s; and widespread interest in them dates from the early 1980s. So, all in all, ANNs deserve our serious attention. The days when they were brushed off as a gimmick or as a mere mental exercise are certainly over. Hybrid ANN/serial computer designs should also be considered to utilize the advantages of both designs where appropriate.

Chapter 2

Fundamentals of Biological Neural Networks

The biological neural network consists of nerve cells (neurons) as in Fig. 2.1, which are interconnected as in Fig. 2.2. The cell body of the neuron, which includes the neuron's nucleus is where most of the neural "computation" takes place. Neural

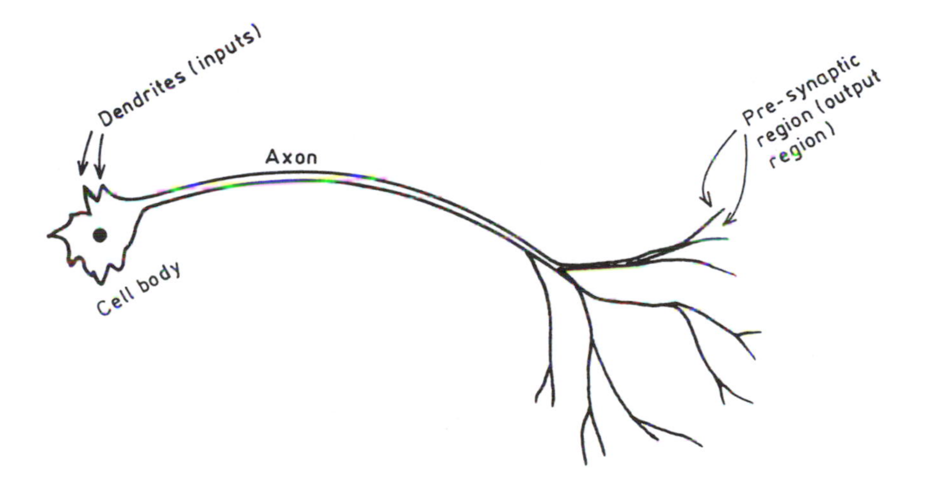

Fig. 2.1. A biological neural cell (neuron).

activity passes from one neuron to another in terms of electrical triggers which travel from one cell to the other down the neuron's axon, by means of an electrochemical process of voltage-gated ion exchange along the axon and of diffusion of neurotransmitter molecules through the membrane over the synaptic gap (Fig. 2.3). The axon can be viewed as a connection wire. However, the mechanism of signal flow is not via electrical conduction but via charge exchange that is transported by diffusion of ions. This transportation process moves along the neuron's cell, down the axon and then through synaptic junctions at the end of the axon via a very narrow synaptic space to the dendrites and/or soma of the next neuron at an average rate of 3 m/sec., as in Fig. 2.3.

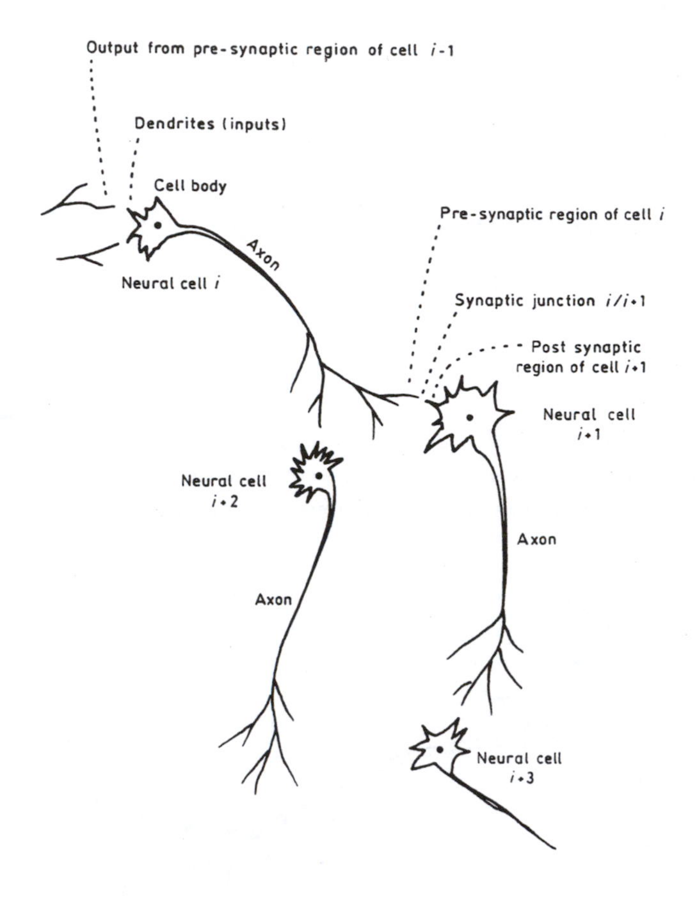

Fig. 2.2. Interconnection of biological neural nets.

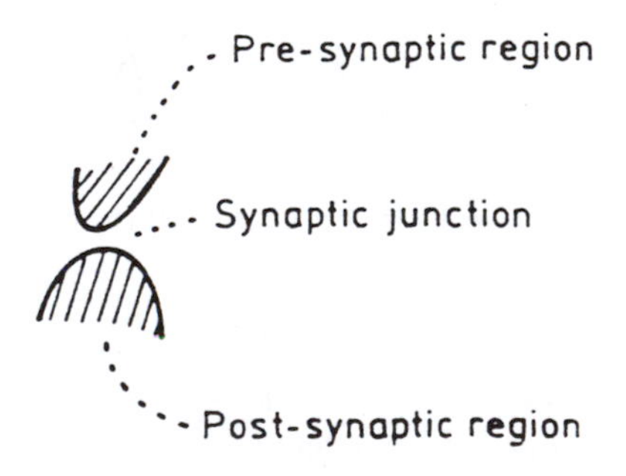

Fig. 2.3. Synaptic junction — detail (of Fig. 2.2).

Figures 2.1 and 2.2 indicate that since a given neuron may have several (hundreds of) synapses, a neuron can connect (pass its message/signal) to many (hundreds of) other neurons. Similarly, since there are many dendrites per each neuron, a single neuron can receive messages (neural signals) from many other neurons. In this manner, the biological neural network interconnects [Ganong, 1973].

It is important to note that not all interconnections, are equally weighted. Some have a higher priority (a higher weight) than others. Also some are excitory and some are inhibitory (serving to block transmission of a message). These differences are effected by differences in chemistry and by the existence of chemical transmitter and modulating substances inside and near the neurons, the axons and in the synaptic junction. This nature of interconnection between neurons and weighting of messages is also fundamental to artificial neural networks (ANNs).

A simple analog of the neural element of Fig. 2.1 is as in Fig. 2.4. In that analog, which is the common building block (neuron) of every artificial neural network, we observe the differences in weighting of messages at the various interconnections (synapses) as mentioned above. Analogs of cell body, dendrite, axon and synaptic junction of the biological neuron of Fig. 2.1 are indicated in the appropriate parts of Fig. 2.4. The biological network of Fig. 2.2 thus becomes the network of Fig. 2.5.

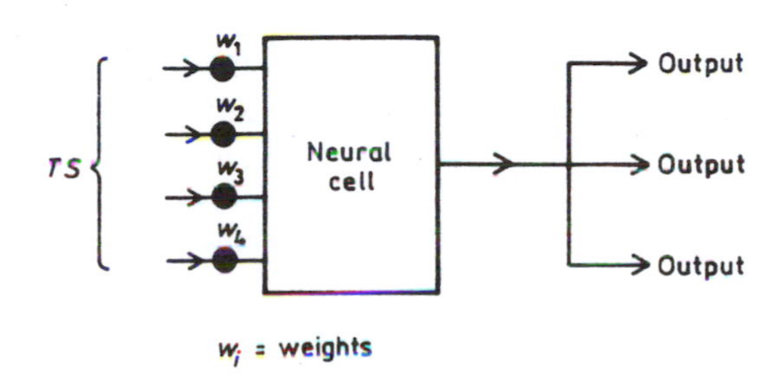

Fig. 2.4. Schematic analog of a biological neural cell.

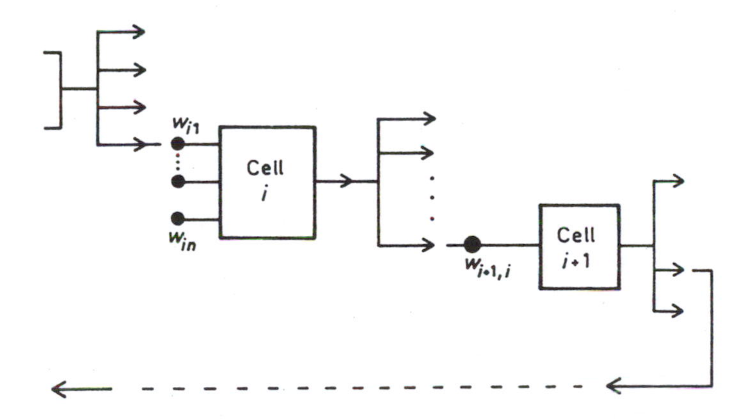

Fig. 2.5. Schematic analog of a biological neural network.

The details of the diffusion process and of charge* (signal) propagation along the axon are well documented elsewhere [B. Katz, 1966]. These are beyond the scope of this text and do not affect the design or the understanding of artificial neural networks, where electrical conduction takes place rather than diffusion of positive and negative ions.

This difference also accounts for the slowness of biological neural networks, where signals travel at velocities of 1.5 to 5.0 meters per second, rather than the speeds of electrical conduction in wires (of the order of speed of light). We comment that discrete digital processing in digitally simulated or realized artificial networks, brings the speed down. It will still be well above the biological networks's speed and is a function of the (micro-) computer instruction execution speed.

*Actually, "charge" does not propagate; membrane polarization change does and is mediated by ionic shifts.

Chapter 3

Basic Principles of ANNs and Their Early Structures

3.1. Basic Principles of ANN Design

The basic principles of the artificial neural networks (ANNs) were first formulated by McCulloch and Pitts in 1943, in terms of five assumptions, as follows:

(1) The activity of a neuron (ANN) is all-or-nothing.
(2) A certain fixed number of synapses larger than 1 must be excited within a given interval of neural addition for a neuron to be excited.
(3) The only significant delay within the neural system is the synaptic delay.
(4) The activity of *any* inhibitory synapse *absolutely* prevents the excitation of the neuron at that time.
(5) The structure of the interconnection network does not change over time.

By assumption (1) above, the neuron is a binary element.

Whereas these are probably historically the earliest systematic principles, they do not all apply to today's state-of-the-art of ANN design.

The *Hebbian Learning Law* (Hebbian Rule) due to Donald Hebb (1949) is also a widely applied principle. The Hebbian Learning Law states that:

"When an axon of cell A is near-enough to excite cell B and when it repeatedly and persistently takes part in firing it, then some growth process or metabolic change takes place in one or both these cells such that the efficiency of cell A [Hebb, 1949] is increased" (i.e. — the weight of the contribution of the output of cell A to the above firing of cell B is increased).

The Hebbian rule can be explained in terms of the following example: Suppose that cell S causes salivation and is excited by cell F which, in turn, is excited by the sight of food. Also, suppose that cell L, which is excited by hearing a bell ring, connects to cell S but cannot alone cause S to fire.

Now, after repeated firing of S by cell F while also cell L is firing, then L will eventually be able to cause S to fire without having cell F fire. This will be due to the eventual increase in the weight of the input from cell L into cell S. Here cells L and S play the role of cells A, B respectively, as in the formulation of the Hebbian rule above.

Also the Hebbian rule need not be employed in all ANN designs. Still, it is implicitly used in designs such as in Chapters 8, 10 and 13.

However, the employment of weights at the input to any neuron of an ANN, and the variation of these weights according to some procedure is common to all ANNs. It takes place in all biological neurons. In the latter, weights variation takes place through complex biochemical processes at the dendrite side of the neural cell, at the synaptic junction, and in the biochemical structures of the chemical messengers that pass through that junction. It is also influenced by other biochemical changes outside the cell's membrane in close proximity to the membrane.

3.2. Basic Network Structures

(1) Historically, the earliest ANNs are *The Perceptron*, proposed by the psychologist Frank Rosenblatt (Psychological Review, 1958).
(2) The Artron (Statistical Switch-based ANN) due to R. Lee (1950s).
(3) The Adaline (Adaptive Linear Neuron, due to B. Widrow, 1960). This artificial neuron is also known as the ALC (adaptive linear combiner), the ALC being its principal component. It is a *single* neuron, not a network.
(4) The Madaline (Many Adaline), also due to Widrow (1988). This is an ANN (network) formulation based on the Adaline above.

Principles of the above four neurons, especially of the Perceptron, are common building blocks in most later ANN developments.

Three later fundamental networks are:

(5) The Back-Propagation network — A multi-layer Perceptron-based ANN, giving an elegant solution to hidden-layers learning [Rumelhart *et al.*, 1986 and others].
(6) The Hopfield Network, due to John Hopfield (1982).

    This network is different from the earlier four ANNs in many important aspects, especially in its recurrent feature of feedback between neurons. Hence, although several of its principles have not been incorporated in ANNs based on the earlier four ANNs, it is to a great extent an ANN-class in itself.
(7) The Counter-Propagation Network [Hecht-Nielsen, 1987] — where Kohonen's Self-Organizing Mapping (SOM) is utilized to facilitate unsupervised learning (absence of a "teacher").

The other networks, such as those of Chaps. 9 to 13 below (ART, Cognitron, LAMSTAR, etc.) incorporate certain elements of these fundamental networks, or use them as building blocks, usually when combined with other decision elements, statistical or deterministic and with higher-level controllers.

3.3. The Perceptron's Input-Output Principles

The Perceptron, which is historically possibly the earliest artificial neuron that was proposed [Rosenblatt, 1958], is also the basic building block of nearly all ANNs. The Artron may share the claim for the oldest artificial neuron. However, it lacks the generality of the Perceptron and of its closely related Adaline, and it was not as influential in the later history of ANN except in its introduction of the statistical switch. Its discussion follows in Sec. 5 below. Here, it suffices to say that its basic structure is as in Fig. 2.5 of Sec. 2, namely, it is a very gross but simple model of the biological neuron, as repeated in Fig. 3.1 below. It obeys the input/output relations

$$Z = \sum_i w_i x_i \tag{3.1}$$

$$y = f_N(z) \tag{3.2}$$

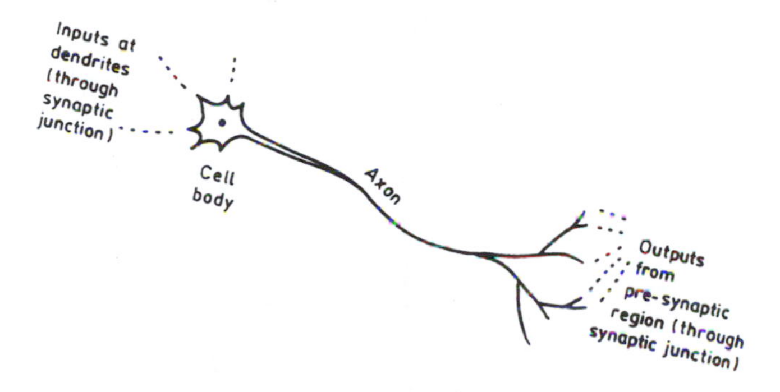

Fig. 3.1. A biological neuron's input output structure. Comment: Weights of inputs are determined through dendritic biochemistry changes and synapse modification. See: M. F. Bear, L. N. Cooper and F. E. Ebner, "A physiological basis for a theory of synapse modification, Science, **237** (1987) 42–48.

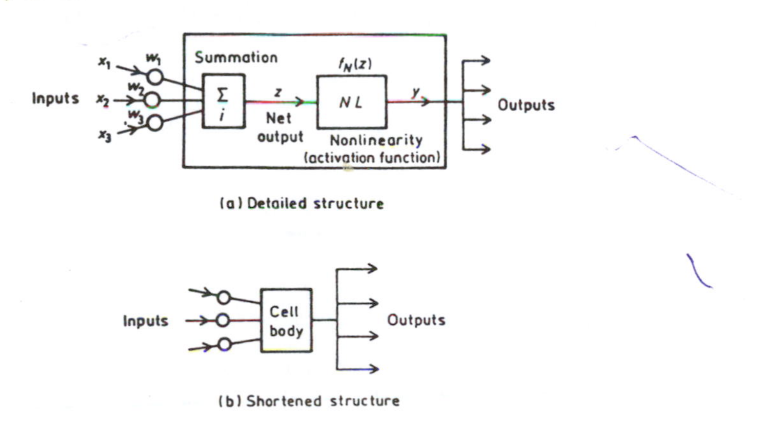

Fig. 3.2. A perceptron's schematic input/output structure.

where w_i is the weight at the inputs x_i where z is the node (summation) output and f_N is a nonlinear operator to be discussed later, to yield the neuron's output y as in Fig. 3.2 is a nonlinear operator to be discussed later, to yield the neuron's output y as in Fig. 3.2.

3.4. The Adaline (ALC)

The Adaline (ADaptive LInear NEuron) of B. Widow (1960) has the basic structure of a bipolar Perceptron as in Sec. 3.1 above and involves some kind of least-error-square (LS) weight training. It obeys the input/node relationships where:

$$z = w_o + \sum_{i=1}^{n} w_i x_i \tag{3.3}$$

where w_o is a bias term and is subject to the training procedure of Sec. 3.4.1 or 3.4.2 below. The nonlinear element (operator) of Eq. (3.2) is here a simple threshold element, to yield the Adaline output y as:

$$y = \text{sign}(z) \tag{3.4}$$

as in Fig. 3.3, such that, for

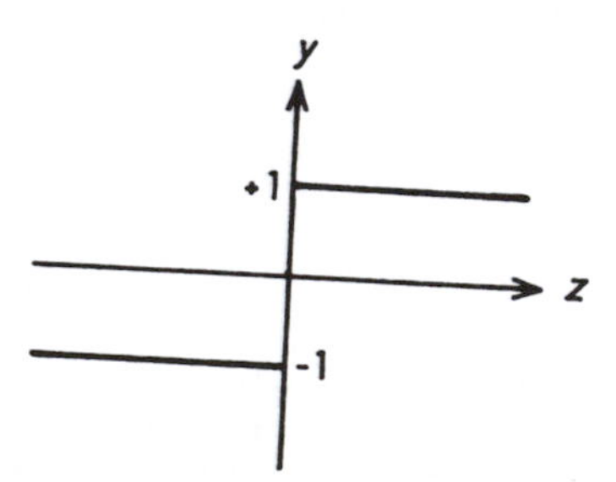

Fig. 3.3. Activation function nonlinearity (Signum function).

$$w_o = 0 \tag{3.5-a}$$

we obtain that

$$z = \sum_i w_i x_i \tag{3.5-b}$$

3.4.1. *LMS training of ALC*

The training of an ANN is the procedure of setting its weights. The training of the Adaline involves training the weights of the ALC (Adaptive Linear Combiner) which is the linear summation element in common to all Adaline/Perceptron neurons. This training is according to the following procedure:

Given L training sets $\mathbf{x}_1 \cdots \mathbf{x}_L$; $\quad d_1 \cdots d_L$
where

$$\mathbf{x}_i = [x_1 \cdots x_n]_i^T \, ; \quad i = 1, 2, \ldots, L \tag{3.6}$$

i denoting the ith set, n being the number of inputs, and d_i denoting the desired outputs of the neuron, we define a training cost, such that:

$$J(\mathbf{w}) \triangleq E\left[e_k^2\right] \cong \frac{1}{L} \sum_{k=1}^{L} e_k^2 \tag{3.7}$$

$$\mathbf{w} \triangleq [w_1 \cdots w_n]_L^T \tag{3.8}$$

E denoting expectation and e_k being a training error at the kth set, namely

$$e_k \triangleq d_k - z_k \tag{3.9}$$

z_k denoting the neuron's actual output.

Following the above notation we have that

$$E\left[e_k^2\right] = E\left[d_k^2\right] + \mathbf{w}^T E\left[\mathbf{x}_k \mathbf{x}_k^T\right] \mathbf{w} - 2\mathbf{w}^T E[d_k \mathbf{x}_k] \tag{3.10}$$

with

$$E[\mathbf{x}\mathbf{x}^T] \triangleq R \tag{3.11}$$

$$E[d\mathbf{x}] = \mathbf{p} \tag{3.12}$$

to yield the gradient ∇J such that:

$$\nabla J = \frac{\partial J(\mathbf{w})}{\partial \mathbf{w}} = 2R\mathbf{w} - 2\mathbf{p} \tag{3.13}$$

Hence, the (optimal) LMS (least mean square) setting of $\mathbf{w}$, namely the setting to yield a minimum cost $J(\mathbf{w})$ becomes:

$$\nabla J = \frac{\partial J}{\partial \mathbf{w}} = 0 \tag{3.14}$$

which, by Eq. (3.13) satisfies the weight setting of

$$\mathbf{w}^{LMS} = R^{-1}\mathbf{p} \tag{3.15}$$

The above LMS procedure employs expecting whereas the training data is limited to a small number of L sets, such that sample averages will be inaccurate estimates of the true expectations employed in the LMS procedure, convergence to the true estimate requiring $L \to \infty$. An alternative to employing small-sample averages of L sets, is provided by using a Steepest Descent (gradient least squares) training procedure for ALC, as in Sec. 3.4.2.

3.4.2. *Steepest descent training of ALC*

The steepest descent procedure for training an ALC neuron does not overcome the shortcomings of small sample averaging, as discussed in relation to the LMS procedure of Sec. 3.4.1 above. It does however attempt to provide weight-setting estimates from one training set to the next, starting estimates from one training set to the next, starting with $L = n + 1$, where n is the number of inputs, noting that to from n weights, it is imperative that

$$L > n + 1 \tag{3.16}$$

The steepest descent procedure, which is a gradient search procedure, is as follows:

Denoting a weights vector setting after the w'th iteration (the m'th training set) as $\mathbf{w}(m)$, then

$$\mathbf{w}(m+1) = \mathbf{w}(m) + \Delta\mathbf{w}(m) \tag{3.17}$$

where $\Delta\mathbf{w}$ is the change (variation) in $\mathbf{w}(m)$, this variation being given by:

$$\Delta\mathbf{w}(m) = \mu \nabla J_{\mathbf{w}(m)} \tag{3.18}$$

μ is the rate parameter whose setting discussed below, and

$$\nabla J = \left[\frac{\partial J}{\partial w_1} \cdots \frac{\partial J}{\partial w_n}\right]^T \tag{3.19}$$

The steepest descent procedure to update $w(m)$ of Eq. (3.17) follows the steps:

(1) Apply input vector $\mathbf{x}_m$ and the desired output d_m for the mth training set.
(2) Determine e_m^2 where

$$\begin{aligned} e_m^2 &= [d_m - \mathbf{w}_{(m)}^T \mathbf{x}(m)]^2 \\ &= d_m^2 - 2d_m \mathbf{w}^T(m)\mathbf{x}(m) + \mathbf{w}^T(m)\mathbf{x}(m)\mathbf{x}^T(m)\mathbf{w}(m) \end{aligned} \tag{3.20}$$

(3) Evaluate

$$\begin{aligned} \nabla J &= \frac{\partial e_m^2}{\partial \mathbf{w}_{(m)}} = 2\mathbf{x}(m)\mathbf{w}^T(m)\mathbf{x}(m) - 2d_m\mathbf{x}(m) \\ &= -2\big[d(m) - \mathbf{w}^T(m)\mathbf{x}(m)\big]\mathbf{x}(m) = -2e_m\mathbf{x}(m) \end{aligned} \tag{3.21}$$

thus obtaining an approximation to ΔJ by using e_m^2 as the approximate to J, namely

$$\nabla J \cong -2e_m x(m)$$

(4) Update $\mathbf{w}(m+1)$ via Eqs. (3.17), (3.18) above, namely

$$\mathbf{w}(m+1) = \mathbf{w}(m) - 2\mu e_m \mathbf{x}(m) \tag{3.22}$$

This is called the *Delta Rule* of ANN.
Here μ is chosen to satisfy

$$\frac{1}{\lambda_{\max}} > \mu > 0 \tag{3.23}$$

if the statistics of x are known, where

$$\lambda_{\max} = \max[\lambda(R)] \tag{3.24}$$

$\lambda(R)$ being an eigenvalve of R of Eq. (3.11) above. Else, one may consider the Droretzky theorem of stochastic approximation [Graupe, *Time Series Anal.*, Chap. 7] for selecting μ, such that

$$\mu = \frac{\mu_0}{m} \tag{3.25}$$

with some convenient μ_0, say $\mu_0 = 1$, to guarantee convergence of $w(m)$ to the unknown but true $\mathbf{w}$ for $m \to \infty$, namely, in the (impractical but theoretical) limit.

Chapter 4

The Perceptron

4.1. The Basic Structure

The Perceptron, which is possibly the earliest neural computation model, is due to F. Rosenblatt and dates back to 1958 (see Sec. 3.1). We can consider the neuronal model using the signum nonlinearity, as in Sec. 3.4) to be a special case of the Perceptron. The Perceptron serves as a building block to most later models, including the Adaline discussed earlier whose neuronal model may be considered as a special case of the Perceptron. The Perceptrron possesses the fundamental structure as in Fig. 4.1 of a neural cell, of several weighted input

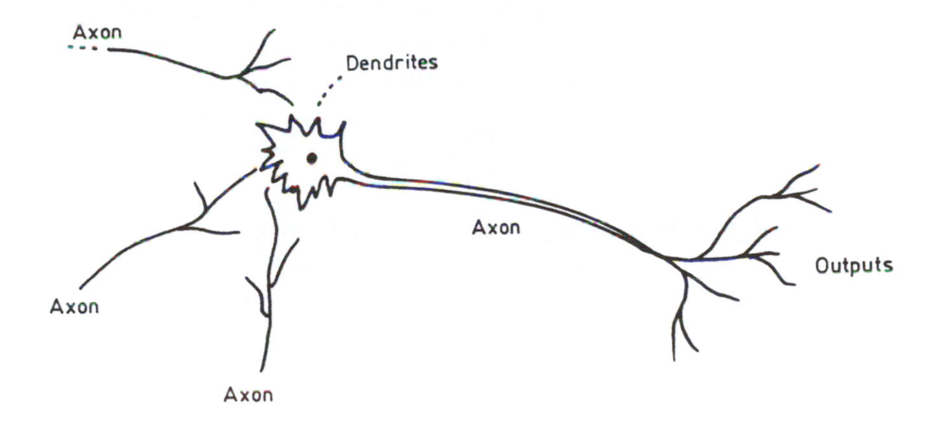

Fig. 4.1. A biological neuron.

connections which connect to the outputs, of several neurons on the input side and of a cell's output connecting to several other neural cells at the output side. It differs from the neuronal model of the Adaline (and Madaline) in its employment of a smooth activation function ("smooth switch" nonlinearity). However the "hard switch" activation function of the Adaline and of the Madaline may be considered as a limit-case of the Perceptron's activation function. The neuronal model of the unit of several weighted inputs/cell/outputs is the perceptron, and it resembles in

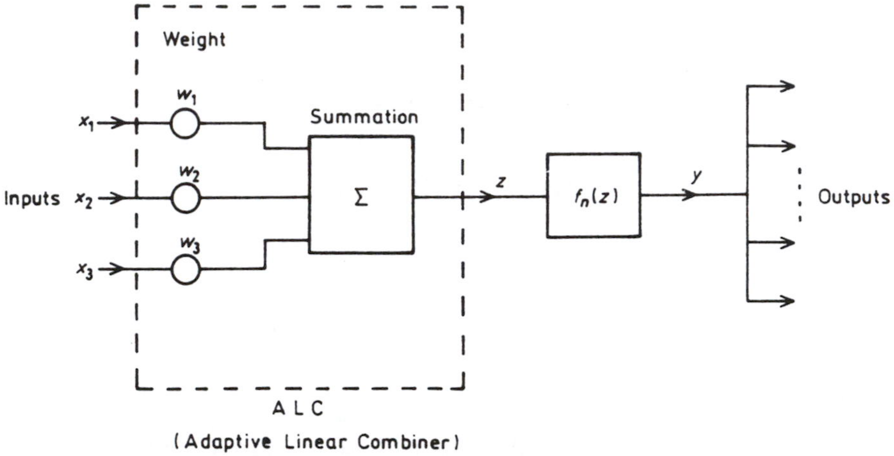

Fig. 4.2. A perceptron (artificial neuron).

structure, in its weighted inputs whose weights are adjustable and in its provision for an output that is a function of the above weighted input, the biological neuron as in Fig. 4.2.

A network of such Perceptrons is thus termed a neural network of Perceptrons. Denoting the summation output of the ith Perceptron as z_i and its inputs as $x_{li} \cdots x_{ni}$, the Perceptron's summation relation is given by

$$z_i = \sum_{j=1}^{m} w_{ij} x_{ij} \tag{4.1}$$

w_{ij} being the weight (which are adjustable as shown below) of the jth input to the ith cell. Equation (4.1) can be written in vector form as:

$$z_i = \mathbf{w}_i^T \mathbf{x}_i \tag{4.2}$$

where

$$\mathbf{w}_i = [w_{i1} \cdots w_{in}]^T \tag{4.3}$$

$$\mathbf{x}_i = [x_{i1} \cdots x_{in}]^T \tag{4.4}$$

T being denoting the transpose of $\mathbf{w}$.

4.1.1. *Perceptron's activation functions*

The Perceptron's cell's output differs from the summation output of Eqs. (4.1) or (4.2) above by the activation operation of the cell's body, just as the output of the biological cell differs from the weighted sum of its input. The

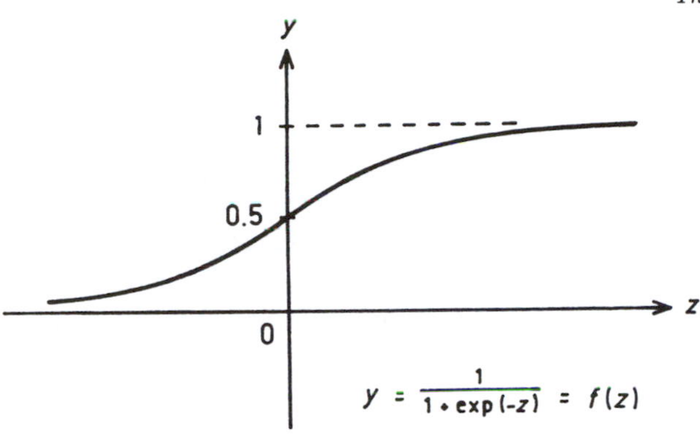

Fig. 4.3. A unipolar activation function for a perceptron.

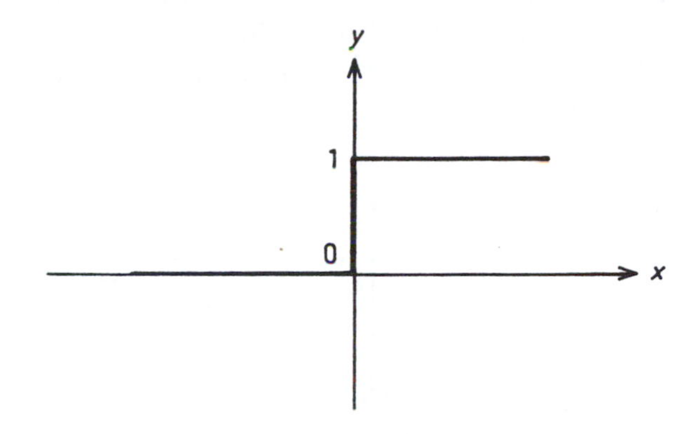

Fig. 4.4. A binary (0,1) activation function.

activation operation is in terms of an *activation function* $f(z_i)$, which is a nonlinear function yielding the ith cell's output y_i to satisfy

$$y_i = f(z_i) \tag{4.5}$$

The activation function f is also known as a squashing function. It keeps the cell's output between certain limits as is the case in the biological neuron. Different functions $f(z_i)$ are in use, all of which have the above limiting property. The most common activation function is the *sigmoid function* which is a continuously differentiable function that satisfies the relation (see Fig. 4.3), as follows:

$$y_i = \frac{1}{1 + \exp(-z_i)} = f(z_i) \tag{4.6}$$

such that for

$$\{z_i \to -\infty\} \Leftrightarrow \{y_i \to 0\}; \ \{z_i = 0\} \Leftrightarrow \{y_i = 0.5\}; \ \{z_i \to \infty\} \Leftrightarrow \{y_i \to 1\}$$

See Fig. 4.4.

Another popular activation function is:

$$y_i = \frac{1 + \tanh(z_i)}{2} = f(z_i) = \frac{1}{1 - \exp(-2z_i)} \tag{4.7}$$

whose shape is rather similar to that of the S-shaped sigmoid function of Eq. (4.6), with $\{z_i \to -\infty\} \Leftrightarrow \{y_i \to 0\}$; $\{z_i = 0\} \Leftrightarrow \{y_i = 0.5\}$ and $\{z_i \to \infty\} \Leftrightarrow \{y_i \to 1\}$

The simplest activation function is a hard-switch limits threshold element; such that:

$$y_i = \begin{cases} 1 & \text{for} \quad z_i \geq 0 \\ 0 & \text{for} \quad z_i < 0 \end{cases} \tag{4.8}$$

as in Fig. 4.4 and as used in the Adaline described earlier (Chap. 4 above). One may thus consider the activation functions of Eqs. (4.6) or (4.7) to be modified binary threshold elements as in Eq. (4.8) where transition when passing through the threshold is being smoothed.

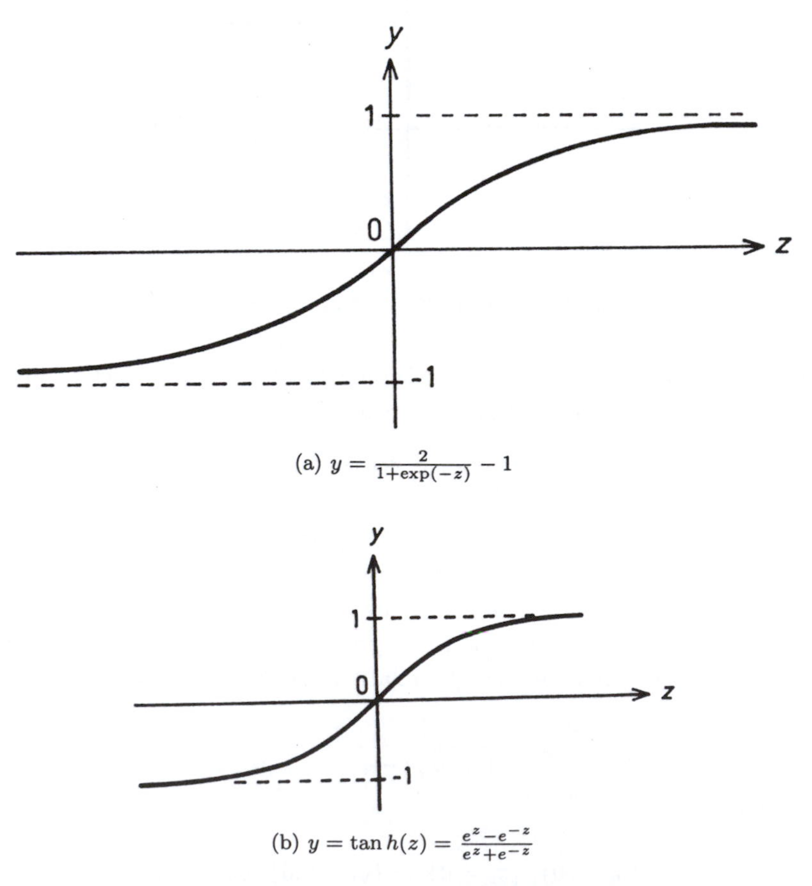

(a) $y = \frac{2}{1+\exp(-z)} - 1$

(b) $y = \tan h(z) = \frac{e^z - e^{-z}}{e^z + e^{-z}}$

Fig. 4.5. Bipolar activation functions.

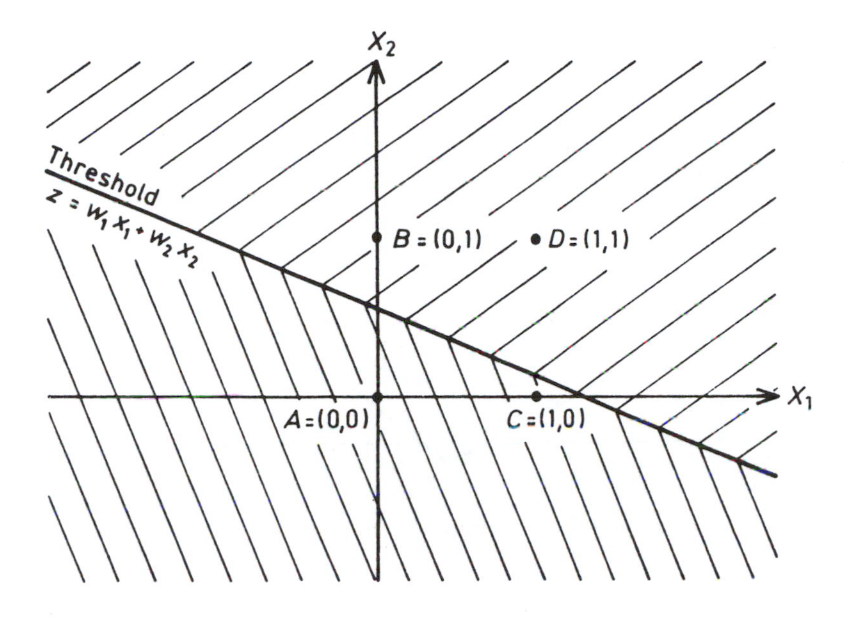

(a) Single-layer perceptron: 2-input representation

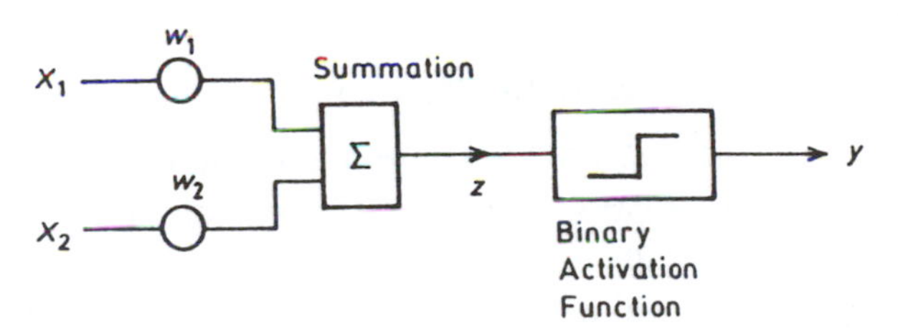

(b) Two-input perceptron

Fig. 4.6. Two-input perceptron and its representation.

In many applications the activation function is moved such that its output y: ranges is from -1 to $+1$ as in Fig. 4.5, rather than from 0 to 1. This is afforded by multiplying the earlier activation function of Eqs. (4.6) or (4.7) by 2 and then subtracting 1.0 from the result, namely, via Eq. (4.6):

$$y_i = \frac{2}{1 + \exp(-z_i)} - 1 = \tanh(z_i/2) \tag{4.9}$$

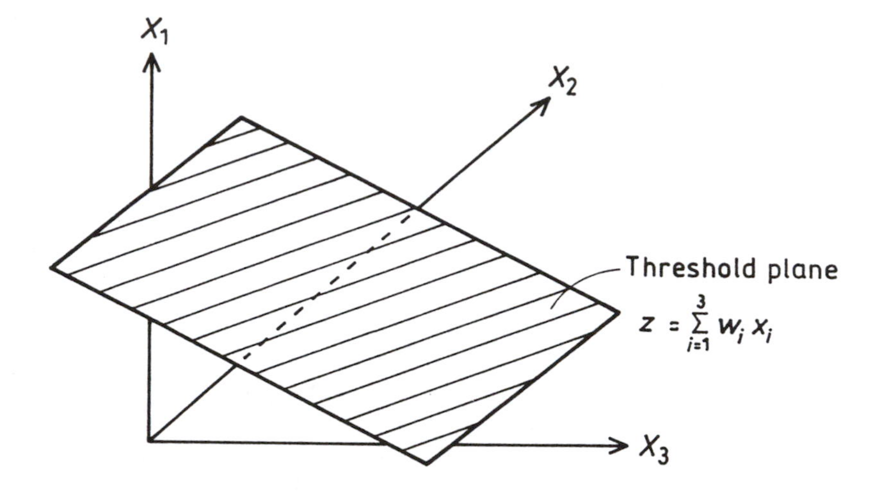

Fig. 4.7. A single layer's 3-input representation.

or, via Eq. (4.7),

$$y_i = \tanh(z_i) = \frac{1 - \exp(-2z_i)}{1 + \exp(-2z_i)} \tag{4.10}$$

Although the Perceptron is only a single neuron (at best, a single-layer network), we present in Sec. 4.A below a case study of its ability to solve a simple linear parameter identification problem.

4.2. The Single-Layer Representation Problem

The *perceptron's learning theorem* was formulated by Rosenblatt in 1961. The theorem states that a perceptron can learn (solve) anything it can represent (simulate). However, we shall see that this theorem does not hold for a single Perceptron (or for any neuronal model with a binary or bipolar output, such as in Chapter 3) or for a single layer of such neuronal models. We shall see later that it does hold for models where the neurons are connected in a multi-layer network.

The single layer perceptron yields the representation description as in Fig. 4.6(a) for a two input situation. This representation holds for several such neurons in a single layer if they do not interconnect.

The above representation diagram results from the perceptron's schematic as in Fig. 4.6(b).

The representation of a 3-input perceptron thus becomes as in Fig. 4.7, where the threshold becomes a flat plane.

By the representation theorem, the perceptron can solve all problems that are or can be reduced to a linear separation (classification) problem.

Table 4.1. XOR Truth-Table.

state	inputs x_1	x_2	output z
A	0	0	0
B	1	0	1
C	0	1	1
D	1	1	0

$(x_1$ or $x_2)$ and $(\bar{x}_1$ or $\bar{x}_2)$;
$\bar{x}$ denoting: not (x)

Table 4.2. Number of linearly separable binary problem. (based on P. P. Wasserman: Neural Computing Theory and Practice © 1989 International Thomson Computer Press. Reprinted with permission).

No. of inputs n	2^{2^n}	No. of linearly separable problems
1	4	4
2	16	14 (all but XOR, XNOR)
3	256	104
4	65 K	1.9 K
5	4.3×10^9	95 K
.	.	.
.	.	.
.	.	.
$n > 7$	x	$< x^{1/3}$

4.3. The Limitations of the Single-Layer Perceptron

In 1969, Minsky and Papert published a book where they pointed out as did E. B. Crane in 1965 in a less-known book, to the grave limitations in the capabilities of the perceptron, as is evident by its representation theorem. They have shown that, for example, the perceptron cannot solve even a 2-state Exclusive-Or (XOR) problem $[(x_1 \cup x_2) \cap (\bar{x}_1 \cup \bar{x}_2)]$, as illustrated in the Truth-Table of Table 4.1, or its complement, the 2-state contradiction problem (XNOR).

Obviously, no linear separation as in Fig. 4.1 can represent (classify) this problem.

Indeed, there is a large class of problems that single-layer classifiers cannot solve. So much so, that for a single layer neural network with an increasing number of inputs, the number of problems that can be classified becomes a very small fraction of the totality of problems that can be formulated.

Specifically, a neuron with binary inputs can have 2^n different input patterns. Since each input pattern can produce 2 different binary outputs, then there are 2^{2^n} different functions of n variables. The number of linearly separable problems of n binary inputs is however a small fraction of 2^{2^n} as is evident from Table 4.2 that is due to Windner (1960). See also Wasserman (1989).

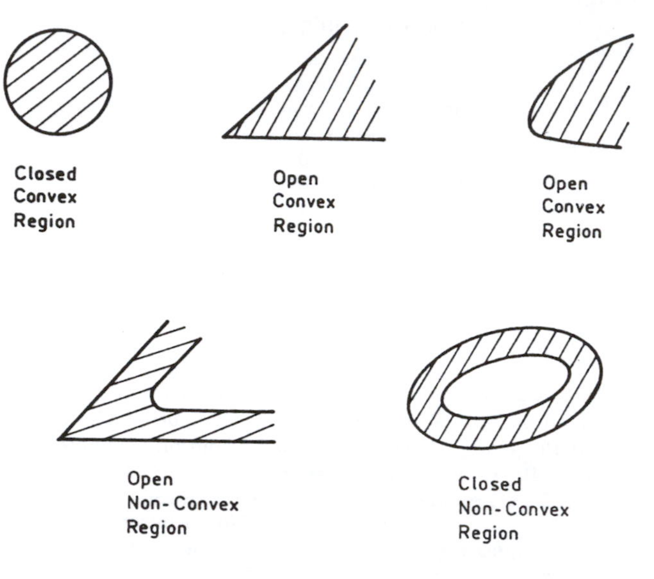

Fig. 4.8. Convex and non-convex regions.

4.4. Many-Layer Perceptrons

To overcome the limitations pointed out by Minsky and Papert, which at the time resulted in a great disappointment with ANNs and in a sharp drop (nearly total) of research into them, it was necessary to go beyond the single layer ANN.

Minsky and Papert (1969) have shown that a single-layer ANN can solve (represent) problems of classification of points that lie in a convex open region or in a convex closed region as in Fig. 4.8. (A convex region is one where any two points in that region can be connected by a straight line that lies fully in that region). In 1969 there was no method to set weights other than for neurons whose output (y) was accessible. It was subsequently shown [Rumelhart *et al.*, 1986] that a 2-layer ANN can solve also non-convex problems, including the XOR problem above. Extension to three or more layers extends the classes of problems that can be represented and hence solved by ANN to, essentially, no bound. However, in the 1960s and 1970s there was no powerful tool to set weights of a multi-layer ANN. Although multilayer training was already used to some extent for the Madaline, it was slow and not rigorous enough for the general multi-layer problem. The solution awaited the formulation of the Back Propagation algorithm, to be described in Chapter 6.

Our comments above, concerning a multi-layer Perceptron network, fully apply to any neuronal model and therefore to any multi-layer neural network, including all networks discussed in later chapters of this text. It therefore applies the Madaline of the next chapter and recurrent networks whose recurrent structure makes a single layer behave as a dynamic multi-layer network.

4.A. Perceptron Case Study: Identifying Autoregressive Parameters of a Signal (AR Time Series Identification)

Goal:

To model a time series parameter identification of a 5th order autoregressive (AR) model using a single Perceptron.

Problem Set Up:

First, a time series signal $x(n)$ of 2000 samples is generated using a *5th order AR model* added with white Gaussian noise $w(n)$. The mathematical model is as follows,

$$x(n) = \sum_{i=1}^{M} a_i x(n-i) + w(n) \tag{4.A.1}$$

where

M = order of the model

a_i = ith element of the AR parameter vector α (alpha)

The true AR parameters as have been used unknown to the neural network to generate the signal $x(u)$, are:

$$\begin{aligned} a_1 &= 1.15 \\ a_2 &= 0.17 \\ a_3 &= -0.34 \\ a_4 &= -0.01 \\ a_5 &= 0.01 \end{aligned}$$

The algorithm presented here is based on deterministic training. A stochastic version of the same algorithm and for the same problem is given in Sec. 11.B below. Given a time series signal $x(n)$, and the order M of the AR model of that signal, we have that

$$\hat{x}(n) = \sum_{i=1}^{M} \hat{a}_i x(n-i) \tag{4.A.2}$$

where $\hat{x}(n)$ is the estimate of $x(n)$, and then define

$$e(n) \triangleq x(n) - \hat{x}(n) \tag{4.A.3}$$

Therefore, if and when $\hat{a}_i$ have converged to a:

$$e(n) \to w(n) \tag{4.A.4}$$

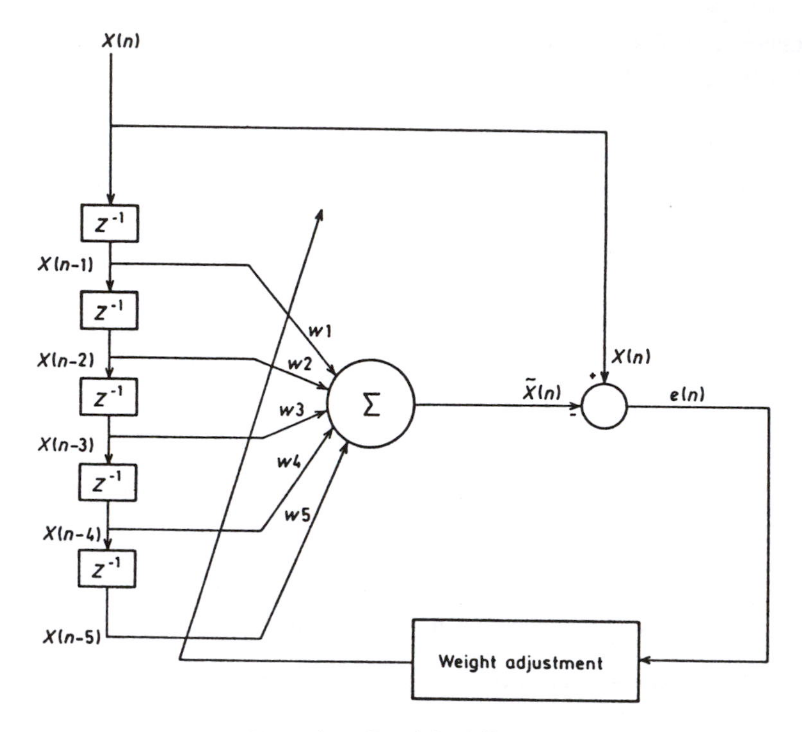

Fig. 4.A.1. Signal flow diagram.

The Perceptron neural network for this model is given in Fig. 4.A.1. Since the white Gaussian noise is uncorrelated with its past,

$$E[w(n)w(n-k)] = \begin{cases} \sigma_x^2 & \text{for } k = 0 \\ 0 & \text{otherwise} \end{cases} \tag{4.A.5}$$

Thus we define a mean square error (MSE) as

$$MSE \triangleq \hat{E}[e^2(n)] = \frac{1}{N}\sum_{i=1}^{N} e^2(i) \tag{4.A.6}$$

which is the sampled variance of the error $e(h)$ above over N samples

Deterministic Training:

Given $x(n)$ from Eq. (4.A.2), find $\hat{a}_i$ such that

$$\hat{x}(n) = \sum_{i=1}^{M} \hat{a}_i x(n-i) = \hat{a}^T x(n-1)$$

$$\hat{\mathbf{a}} \triangleq [\hat{a}_1 \cdots \hat{a}_M]^T$$

then calculate

$$e(n) = x(n) - \hat{x}(n) \tag{4.A.7}$$

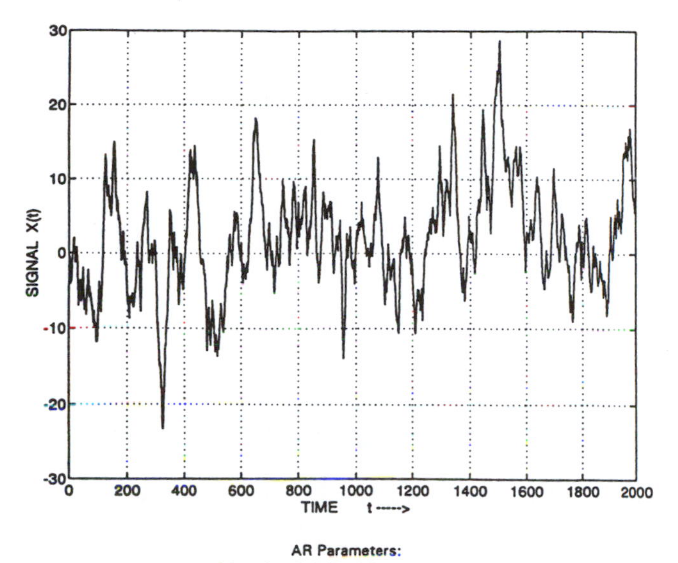

Fig. 4.A.2. Signal versus time.

update the weight vector $\hat{\mathbf{a}}$ to minimize the MSE error of Eq. (4.A.6), by using the delta rule and momentum term

$$\Delta\hat{\mathbf{a}}(n) = 2\mu e(n)\mathbf{x}(n-1) + \alpha\Delta\hat{\mathbf{a}}(n-1) \tag{4.A.8}$$

$$\hat{\mathbf{a}}(n+1) = \hat{\mathbf{a}}(n) + \Delta\hat{\mathbf{a}}(n) \tag{4.A.9}$$

where

$$\hat{\mathbf{a}}(n) = [\hat{a}_1(n), \ldots, \hat{a}_5(n)]^T$$

$$\mathbf{x}(n-1) = [x(n-1)\cdots x(n-5)]^T$$

$$\mu_0 = 0.001$$

$$\alpha = 0.5$$

and μ is decreasing in iteration step as,

$$\mu = \frac{\mu_0}{1+k} \tag{4.A.10}$$

Note that α is a momentum coefficient which is added to the update equation since it can serve to increase the speed of convergence.

A plot of MSE versus the number of iteration is shown in Fig. 4.A.3. The flow chart of deterministic training is shown in Fig. 4.A.4.

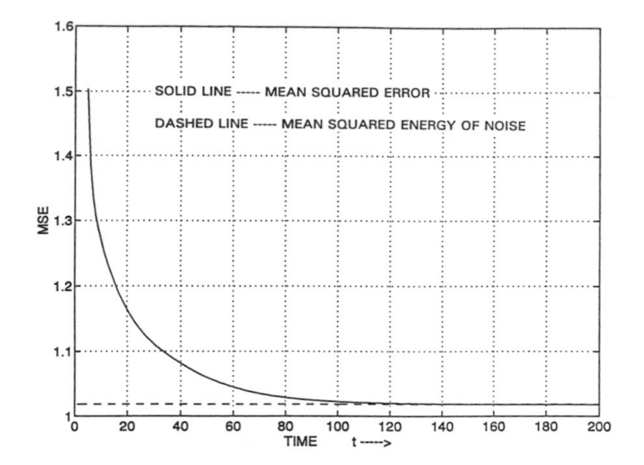

Fig. 4.A.3. Mean squared error versus time.

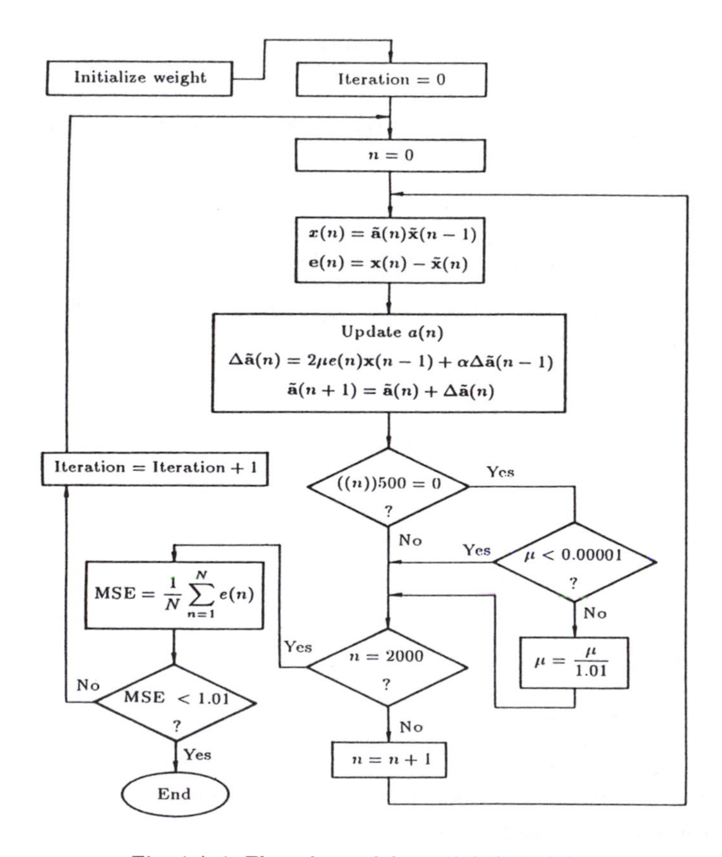

Fig. 4.A.4. Flow chart of deterministic training.

Program Printout: written in MATLAB® (MATLAB is a registered trademark of The MathWorks, Inc.)

```
%%%%%%%%%%%%%%%%%%%%%%%%%%%%%%%%%%%%%%%%%%%%%%%%%%%%%%%%%%%%%%%%%%%%%%%%%%%%%%%
%
%  MATLAB FILE
%
%
%
%
%
%
%%%%%%%%%%%%%%%%%%%%%%%%%%%%%%%%%%%%%%%%%%%%%%%%%%%%%%%%%%%%%%%%%%%%%%%%%%%%%%%

f1 = fopen('ESTIMATEBP1.dat','w');
fprintf(f1,'\nTHE FOLLOWING IS THE WEIGHT CHANGE AND THE MEAN SQUARE ERROR');
fprintf(f1,'\n-------------------------------------------------------------------\n\n');
w1 = rand(5,1)/5;
w2 = rand(1)/5; bias = 4.0; delw2 = 0;
mu = 0.001;
delw1 = 0;  momentum = 0.5;

n = 5; ITERATION=2000;
Xpad = zeros(ITERATION+n,1);
Xpad(n+1:n+ITERATION) = X;
Xest = zeros(ITERATION+n,1);
error = zeros(size(X));
MSE = zeros(200,1);
count = 1;
        for loop=1:400
            totalerr = 0;
            for i=n+1:n+ITERATION
                xt = Xpad(i-n:i-1) ;
                dt = Xpad(i);
                z1 = w1'*xt + w2*bias;
                error(i) = dt - z1;
                phi_o = (dt-z1)*xt;
                delw1 = mu*phi_o + momentum*delw1;
                delw2 = mu*(dt-z1)*bias + momentum*delw2;
                w1 = w1 + delw1;
                w2 = w2 + delw2;
                if round(i/500)*500 == i
                fprintf(f1,'\n\t%6.3f\t%6.3f\t%6.3f\t%6.3f\t%6.3f',fliplr(w1'));
                   if mu > 0.000001
                      mu = mu/1.01;
                   end
                end
            end
            if  round(loop/1)*1 == loop
                MSE(loop) = (error'*error)/ITERATION;
                fprintf(f1,'\n\t%6.3f\t%6.3f\t%6.3f\t%6.3f\t%6.3f',fliplr(w1'));
                fprintf(f1,'\n  %d\t       MEANSQUARE ERROR = %6.4f\n\n',loop,MSE(loop));
                disp([loop      fliplr(w1')])
                momentum = momentum/1.00001;
                if MSE < 0.019
                   loop = 1000
                end
            end
        end
fclose(f1);
```

Computational Results: Parameter Estimates (Weights) and Mean Square Error — Deterministic Training, No Bias Term Added

```
        1.15    0.17   -0.34   -0.01    0.01 = true (unknown) parameters

        0.598   0.446  -0.171   0.380  -0.231
1            MEANSQUARE ERROR = 117.0848

        0.743   0.353  -0.254   0.302  -0.131
2            MEANSQUARE ERROR = 17.5898

        0.808   0.300  -0.247   0.209  -0.062
3            MEANSQUARE ERROR = 4.2322

        0.838   0.271  -0.229   0.145  -0.021
4            MEANSQUARE ERROR = 1.9580

        0.856   0.253  -0.214   0.106   0.002
5            MEANSQUARE ERROR = 1.5033

        0.869   0.240  -0.205   0.082   0.014
6            MEANSQUARE ERROR = 1.3857

        0.881   0.230  -0.201   0.069   0.021
7            MEANSQUARE ERROR = 1.3386

        0.892   0.221  -0.200   0.060   0.026
8            MEANSQUARE ERROR = 1.3103

        0.901   0.214  -0.201   0.054   0.029
9            MEANSQUARE ERROR = 1.2892

        0.911   0.208  -0.203   0.050   0.031
10           MEANSQUARE ERROR = 1.2717

        0.920   0.203  -0.206   0.046   0.032
11           MEANSQUARE ERROR = 1.2564

        0.928   0.199  -0.209   0.042   0.034
12           MEANSQUARE ERROR = 1.2427

        0.936   0.195  -0.212   0.039   0.035
13           MEANSQUARE ERROR = 1.2301

        0.944   0.191  -0.215   0.036   0.036
14           MEANSQUARE ERROR = 1.2185

        0.951   0.188  -0.218   0.034   0.036
15           MEANSQUARE ERROR = 1.2077
```

```
32          MEANSQUARE ERROR = 1.1047

        1.040   0.165  -0.262   0.000   0.036
33          MEANSQUARE ERROR = 1.1014

        1.043   0.165  -0.264  -0.001   0.035
34          MEANSQUARE ERROR = 1.0982

        1.046   0.164  -0.266  -0.003   0.035
35          MEANSQUARE ERROR = 1.0952

        1.050   0.164  -0.268  -0.004   0.035
36          MEANSQUARE ERROR = 1.0923

        1.053   0.163  -0.269  -0.006   0.034
37          MEANSQUARE ERROR = 1.0894

        1.056   0.163  -0.271  -0.007   0.034
38          MEANSQUARE ERROR = 1.0867

        1.059   0.163  -0.273  -0.008   0.033
39          MEANSQUARE ERROR = 1.0840

        1.062   0.162  -0.274  -0.009   0.033
40          MEANSQUARE ERROR = 1.0814

        1.064   0.162  -0.276  -0.011   0.033
41          MEANSQUARE ERROR = 1.0789

        1.067   0.162  -0.277  -0.012   0.032
42          MEANSQUARE ERROR = 1.0765

        1.070   0.162  -0.278  -0.013   0.032
43          MEANSQUARE ERROR = 1.0741

        1.072   0.161  -0.280  -0.014   0.031
44          MEANSQUARE ERROR = 1.0718

        1.075   0.161  -0.281  -0.015   0.031
45          MEANSQUARE ERROR = 1.0696

        1.077   0.161  -0.282  -0.016   0.030
46          MEANSQUARE ERROR = 1.0675

        1.079   0.161  -0.283  -0.018   0.030
47          MEANSQUARE ERROR = 1.0654

        1.081   0.161  -0.284  -0.019   0.030
48          MEANSQUARE ERROR = 1.0634
```

```
          1.137   0.169  -0.302  -0.040   0.019
181            MEANSQUARE ERROR = 1.0181

          1.137   0.169  -0.302  -0.040   0.019
182            MEANSQUARE ERROR = 1.0181

          1.137   0.169  -0.302  -0.040   0.019
183            MEANSQUARE ERROR = 1.0181

          1.137   0.169  -0.302  -0.040   0.019
184            MEANSQUARE ERROR = 1.0181

          1.137   0.169  -0.302  -0.040   0.019
185            MEANSQUARE ERROR = 1.0181

          1.137   0.169  -0.302  -0.040   0.019
186            MEANSQUARE ERROR = 1.0181

          1.137   0.169  -0.302  -0.040   0.019
187            MEANSQUARE ERROR = 1.0181

          1.137   0.169  -0.302  -0.040   0.019
188            MEANSQUARE ERROR = 1.0181

          1.137   0.169  -0.302  -0.040   0.019
189            MEANSQUARE ERROR = 1.0181

          1.137   0.169  -0.302  -0.040   0.020
190            MEANSQUARE ERROR = 1.0181

          1.137   0.169  -0.302  -0.040   0.020
191            MEANSQUARE ERROR = 1.0181

          1.137   0.169  -0.302  -0.040   0.020
192            MEANSQUARE ERROR = 1.0181

          1.137   0.169  -0.302  -0.040   0.020
193            MEANSQUARE ERROR = 1.0181

          1.137   0.169  -0.302  -0.040   0.020
194            MEANSQUARE ERROR = 1.0181

          1.137   0.169  -0.302  -0.040   0.020
195            MEANSQUARE ERROR = 1.0181

          1.137   0.169  -0.302  -0.040   0.020
196            MEANSQUARE ERROR = 1.0181

          1.137   0.169  -0.302  -0.040   0.020
```

```
197        MEANSQUARE ERROR = 1.0181

      1.137   0.169  -0.302  -0.040   0.020
198        MEANSQUARE ERROR = 1.0181

      1.137   0.169  -0.302  -0.040   0.020
199        MEANSQUARE ERROR = 1.0181

      1.137   0.169  -0.302  -0.040   0.020
200        MEANSQUARE ERROR = 1.0181
```

Parameter Estimates (Weights) and The Mean Square Error Deterministic Training only with Bias Term Added

	1.15	0.17	-0.34	-0.01	0.01	= true (unknown) parameters
1	0.587	0.451	-0.180	0.380	-0.246	MEANSQUARE ERROR = 122.0708
2	0.733	0.353	-0.257	0.303	-0.147	MEANSQUARE ERROR = 18.3725
3	0.800	0.297	-0.249	0.209	-0.076	MEANSQUARE ERROR = 4.3699
4	0.831	0.268	-0.230	0.145	-0.035	MEANSQUARE ERROR = 1.9879
5	0.850	0.249	-0.215	0.105	-0.011	MEANSQUARE ERROR = 1.5141
6	0.863	0.236	-0.205	0.081	0.002	MEANSQUARE ERROR = 1.3927
7	0.875	0.226	-0.201	0.067	0.009	MEANSQUARE ERROR = 1.3445
8	0.886	0.217	-0.200	0.058	0.014	MEANSQUARE ERROR = 1.3158
9	0.896	0.210	-0.201	0.052	0.017	MEANSQUARE ERROR = 1.2943
10	0.906	0.204	-0.203	0.048	0.020	MEANSQUARE ERROR = 1.2765
11	0.915	0.199	-0.206	0.044	0.022	MEANSQUARE ERROR = 1.2609
12	0.924	0.195	-0.209	0.040	0.023	MEANSQUARE ERROR = 1.2469
13	0.932	0.191	-0.212	0.037	0.025	MEANSQUARE ERROR = 1.2341
14	0.940	0.188	-0.215	0.035	0.026	MEANSQUARE ERROR = 1.2223
15	0.947	0.185	-0.218	0.032	0.027	MEANSQUARE ERROR = 1.2113

```
       1.136   0.169  -0.301  -0.040   0.018
181          MEANSQUARE ERROR = 1.0168

       1.136   0.169  -0.301  -0.040   0.018
182          MEANSQUARE ERROR = 1.0168

       1.136   0.169  -0.301  -0.040   0.018
183          MEANSQUARE ERROR = 1.0168

       1.136   0.169  -0.301  -0.040   0.018
184          MEANSQUARE ERROR = 1.0168

       1.136   0.169  -0.301  -0.040   0.018
185          MEANSQUARE ERROR = 1.0168

       1.136   0.169  -0.301  -0.040   0.018
186          MEANSQUARE ERROR = 1.0168

       1.136   0.169  -0.301  -0.040   0.018
187          MEANSQUARE ERROR = 1.0168

       1.136   0.169  -0.301  -0.040   0.018
188          MEANSQUARE ERROR = 1.0168

       1.136   0.169  -0.301  -0.040   0.018
189          MEANSQUARE ERROR = 1.0168

       1.136   0.169  -0.301  -0.040   0.018
190          MEANSQUARE ERROR = 1.0168

       1.136   0.169  -0.302  -0.040   0.018
191          MEANSQUARE ERROR = 1.0168

       1.136   0.169  -0.302  -0.040   0.018
192          MEANSQUARE ERROR = 1.0168

       1.136   0.169  -0.302  -0.040   0.018
193          MEANSQUARE ERROR = 1.0168

       1.136   0.169  -0.302  -0.040   0.018
194          MEANSQUARE ERROR = 1.0168

       1.136   0.169  -0.302  -0.040   0.018
195          MEANSQUARE ERROR = 1.0168

       1.136   0.169  -0.302  -0.040   0.018
196          MEANSQUARE ERROR = 1.0168
```

```
              1.136    0.169   -0.302   -0.040    0.018
197                MEANSQUARE ERROR = 1.0168

              1.136    0.169   -0.302   -0.040    0.018
198                MEANSQUARE ERROR = 1.0168

              1.136    0.169   -0.302   -0.040    0.018
199                MEANSQUARE ERROR = 1.0168

              1.136    0.169   -0.302   -0.040    0.018
200                MEANSQUARE ERROR = 1.0168
```

Observe the closeness of the parameters identified above (say, at iteration 200) to the original but unknown parameters as at the beginning of Sec. 4.A.

Chapter 5

The Madaline

The Madaline (Many Adaline) is a multilayer extension of the single-neuron bipolar Adaline to a network. It is also due to B. Widrow (1988). Since the Madaline network is a direct multi-layer extension of the Adaline of Sec. 3, we present it before discussing the Back Propagation network that is historically earlier (see our discussion in Sec. 4.4 above). Its weight adjustment methodology is more intuitive than in Back Propagation and provides understanding into the difficulty of adjusting weights in a multi-layer network, though it is less efficient. Its basic structure is given in Fig. 5.1 which is in terms of two layers of Adalines, plus an input layer which merely serves as a network's input distributor (see Fig. 5.2).

5.1. Madaline Training

Madaline training differs from Adaline training in that no partial desired outputs of the inside layers are or can be available. The inside layers are thus termed *hidden layers*. Just as in the human central nervous system (CNS), we may receive learning information in terms of desired and undesired outcome, though the human is not conscious of outcomes of individual neurons inside the CNS that participate in that learning, so in ANN no information of inside layers of neurons is available.

The Madaline employs a training procedure known as Madaline Rule II, which is based on a *Minimum Disturbance Principle*, as follows [Widrow *et al.*, 1987]:

(1) All weights are initialized at low random values. Subsequently, a training set of L input vectors $\mathbf{x}_i (i = 1, 2, \ldots, L)$ is applied one vector at a time to the input.
(2) The number of incorrect bipolar values at the output layer is counted and this number is denoted as the error e per a given input vector.
(3) For all neurons at the output layer:
    (a) Denoting *th* as the threshold of the activation function (preferably 0), check: [*z-th*] for *every* input vector of the given training set of vectors for the particular layer that is considered at this step. Select the first unset neuron from the above but which corresponds to the lowest *abs*[*z-th*] occurring over that set of input vectors. Hence, for a case of L input vectors in an input set and for a layer of n neurons, selection is from $n \times L$ values of z. This is

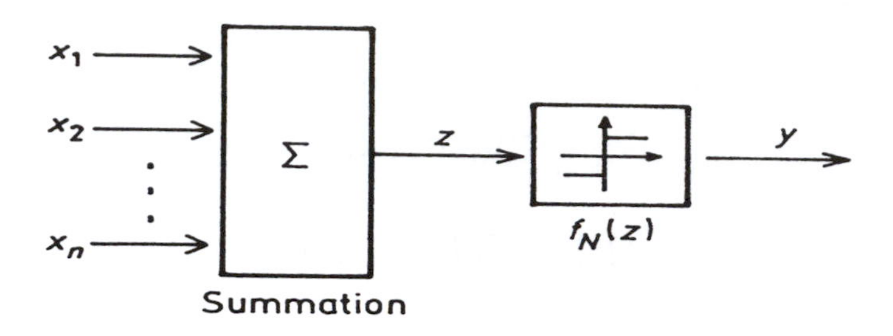

Fig. 5.1. A simple Madaline structure.

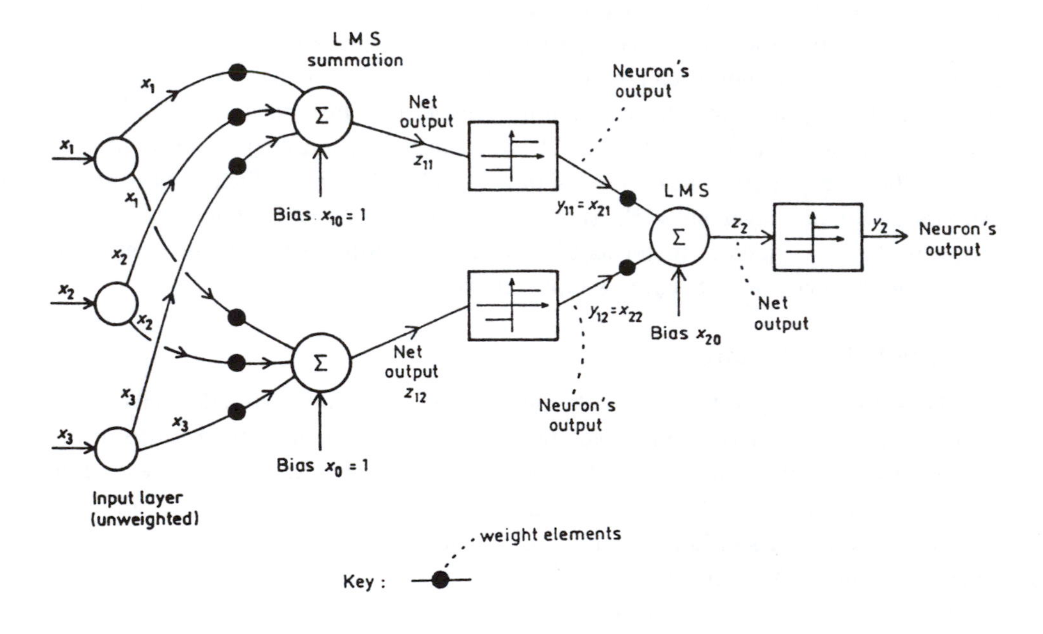

Fig. 5.2. The Madaline network of 2 layers.

the node that can reverse its polarity by the smallest change in its weights, thus being denoted as the *minimum-disturbance neuron*, from which the procedures name is derived. A previously unset neuron is a neuron whose weights have not been set yet.

(b) Subsequently, one should change the weights of the latter neuron such that the bipolar output y of that unit changes. The smallest change in weight via a modified steepest procedure as in Sec. 3.4.2 that considers $[z\text{-}th]$ instead of e_m of Eq. (3.22) will cause this change. Alternatively, random changes may be employed.

(c) The input set of vectors is propagated to the output once again.
(d) If the change in weight reduced the performance cost "e" of Step 2, then this change is accepted. Else, the original (earlier) weights are restored to that neuron.

(4) Repeat Step 3 for all layers except for the input layer.
(5) For all neurons of the output layer: Apply Steps 3, 4 for a pair of neurons whose analog node-outputs z are closest to zero, etc.
(6) For all neurons of the output layer: Apply Steps 3, 4 for a triplet of neurons whose analog node-outputs are closest to zero, etc.
(7) Go to next vector up to the L'th vector.
(8) Repeat for further combinations of L vectors till training is satisfactory.

The same can be repeated for quadruples of neurons, etc. However, this setting then becomes very lengthy and may therefore be unjustified. All weights are initially set to (different) low random values. The values of the weights can be positive or negative within some fixed range, say, between -1 and 1. The initial learning rate μ of Eq. (3.18) of the previous chapter should be between 1 and 20. For adequate convergence, the number of hidden layer neurons should be at least 3, preferably higher. Many iterations steps (often, thousands) of the steepest descent algorithm of Sec. 3.4.2 are needed for convergence. It is preferable to use a bipolar rather than a binary configuration for the activation function.

The above discussion of the Madeline neural network (NN) indicates that the Madeline is an intuitive but rather primitive and inefficient NN. It is also very sensitive to noise. Though it has the basic properties of several other neural networks discussed in later chapters of this text, we shall see that the networks discussed later are considerably more efficient and less noise-sensitive.

5.A. Madaline Case Study*: Character Recognition

5.A.1. *Problem statement*

Designing a Madaline (Multiple Adaline) Neural Network to recognize 3 characters 0, C and F supplied in a binary format and represented using a 6×6 grid. The Neural Network should be trained and tested with various patterns and the total error rate and the amount of convergence should be observed. Typical patterns used for training and testing are as in Fig. 5.A.1.

*Computed by Vasanath Arunachalam, ECS Dept. University of Illinois, Chicago, 2006.

Fig. 5.A.1: *Patterns to be recognized*

1	1	1	1	1	1
1	-1	-1	-1	-1	-1
1	-1	-1	-1	-1	-1
1	-1	-1	-1	-1	-1
1	-1	-1	-1	-1	-1
1	1	1	1	1	1

Fig. 5.A.1(a). Pattern representing character C.

1	1	1	1	1	1
1	-1	-1	-1	-1	1
1	-1	-1	-1	-1	1
1	-1	-1	-1	-1	1
1	-1	-1	-1	-1	1
1	1	1	1	1	1

Fig. 5.A.1(b). Pattern representing character 0.

1	1	1	1	1	1
1	-1	-1	-1	-1	-1
1	1	1	1	1	1
1	-1	-1	-1	-1	-1
1	-1	-1	-1	-1	-1
1	-1	-1	-1	-1	-1

Fig. 5.A.1(c). Pattern representing character F.

5.A.2. *Design of network*

A Madaline network as in Fig. 5.A.2 was implemented with 3 layer, input (6 neurons), hidden (3 neurons), and output (2 neurons), layers. 36 inputs from a grid containing characters 0, C or F are given as input to the network. 15 such input sets are given, 5 each for 3's and 0's. The weights of the network are initially set in a random fashion in the range $\{-1, 1\}$.

Fig. 5.A.2: ***The Madaline network***

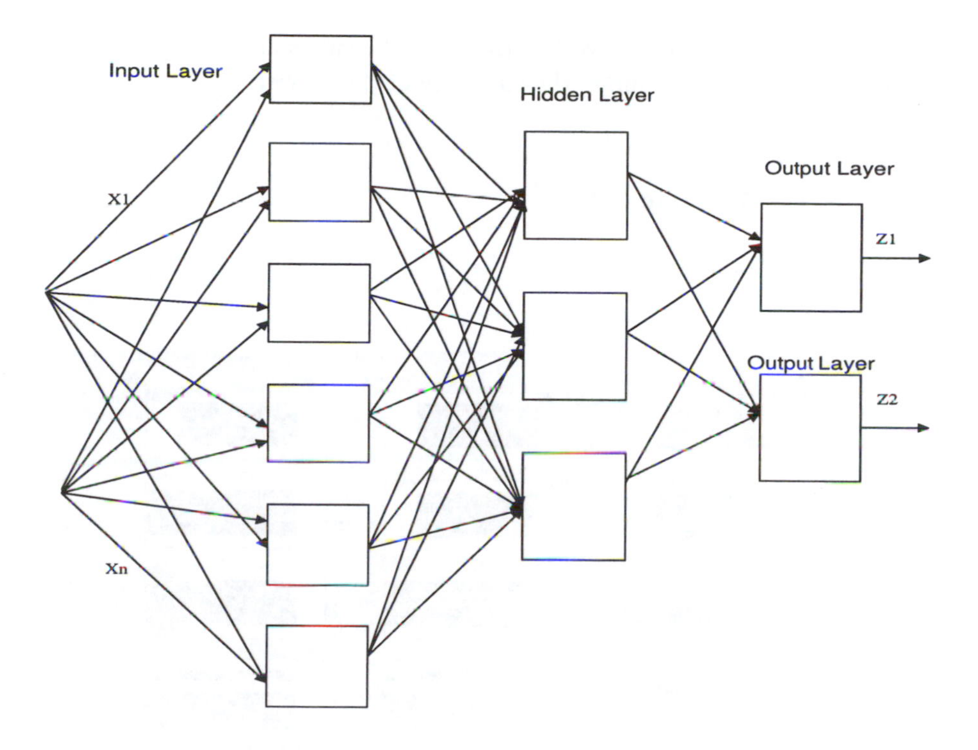

5.A.3. *Training of the network*

The following are the basic steps for Training of a Back Propagation Neural Network

- Generate a training data set with 5 sets of 0's, C's and F's each.
- Feed this training set (see Fig. 5.A.3) to the network.
- Set weights of the network randomly in the range $\{-1, 1\}$.
- Use hardlimiter transfer function for each neuron.

$$Y(n) = \begin{cases} 1, & \text{if } x \geqq 0 \\ -1, & \text{if } x < 0 \end{cases}$$

- Each output is passed as input to the successive layer.
- The final output is compare with the desired output and cumulative error for the 15 inputs is calculated.
- If the error percent is above 15% then the weights (for the neuron which has output closest to 0) of the output layer is changed using

$$\text{weight}_{\text{new}} = \text{weight}_{\text{old}} + 2^*\text{constant}^*\text{output (previous layer)}^*\text{error}$$

- Weight(s) are updated and the new error is determined.
- Weights are updated for various neurons until there is no error or the error is below a desired threshold.
- Test data set is fed to the network with updated weights and the output (error) is obtained thereby determining the efficiency of the network.

Fig. 5.A.3: ***The Training Sets:***

Fig. 5.A.1(a): Training Set 1

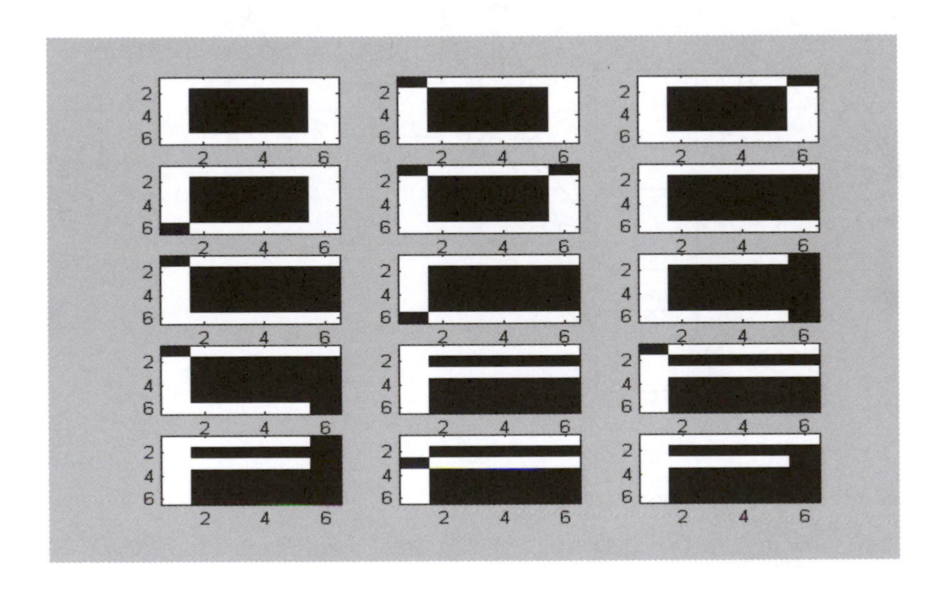

Fig. 5.A.3(b): Test Set 2

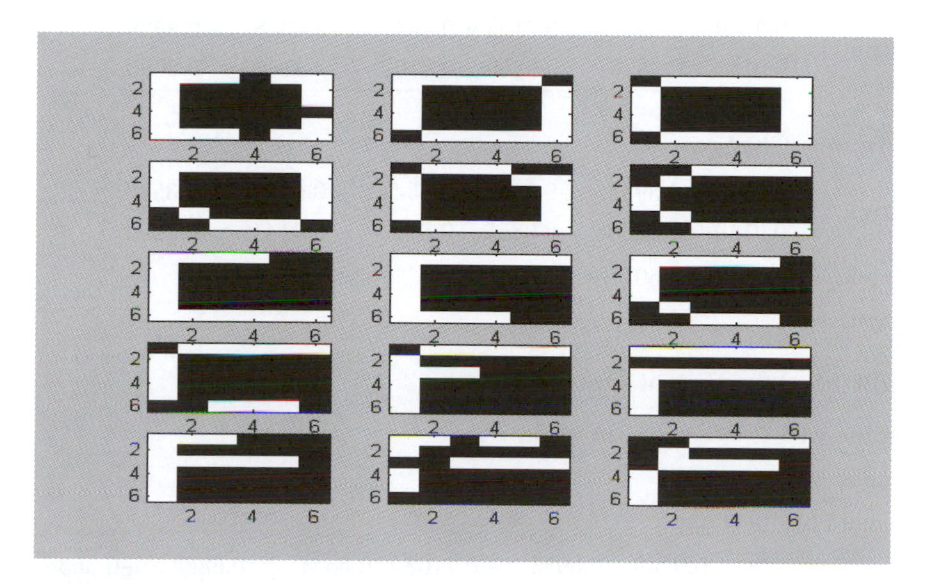

5.A.4. *Results*

The results are as shown below:

- Hidden Layer Weight Matrix:

w_hidden =

Columns 1 through 12

−0.9830	0.6393	0.1550	−0.2982	−0.7469	−0.0668	0.1325	−0.9485	0.2037	0.1573	0.1903	−0.8288
0.2329	−0.1504	0.6761	0.0423	0.6629	0.1875	0.1533	−0.1751	−0.6016	−0.9714	0.7511	−0.3460
0.9927	−0.4033	0.4272	0.8406	0.6071	0.5501	−0.3400	−0.8596	−0.7581	0.3686	−0.6020	−0.6334
0.8494	−0.7395	−0.2944	0.7219	−0.1397	−0.4833	0.5416	−0.8979	−0.1973	0.6348	−0.2891	0.5008
0.7706	0.9166	−0.0775	−0.4108	−0.1773	−0.6749	0.4772	0.1271	−0.8654	0.7380	−0.0697	0.4995
0.5930	−0.0853	0.8175	−0.0605	−0.7407	0.4429	0.6812	−0.7174	0.9599	−0.3352	−0.3762	−0.5934

Columns 13 through 24

0.5423	0.1111	0.7599	−0.4438	−0.5097	0.9520	−0.1713	−0.7768	−0.1371	0.7247	−0.2830	0.4197
−0.2570	−0.4116	−0.3409	0.5087	0.6436	−0.0342	−0.7515	−0.7608	0.2439	−0.8767	0.4824	−0.3426
0.6383	−0.0592	0.9073	0.0101	−0.2051	0.9051	−0.6792	0.4301	−0.7850	−0.1500	−0.2993	0.2404
−0.2520	0.2275	0.1467	0.3491	−0.5696	−0.7650	−0.3104	0.5042	−0.8040	0.5050	0.1335	0.1340
−0.0943	0.9710	−0.2042	−0.6193	−0.8348	0.3316	0.4818	−0.7792	0.6217	0.9533	0.3451	0.7745
−0.2432	−0.1404	−0.7061	−0.8046	−0.6752	0.6320	−0.2957	0.9080	0.5916	−0.7896	0.6390	0.4778

Columns 25 through 36

0.1716	−0.2363	0.8769	0.6879	0.6093	−0.3614	−0.6604	−0.6515	0.4398	0.4617	−0.8053	0.5862
0.7573	−0.4263	−0.6195	−0.4669	0.1387	−0.0657	−0.6288	−0.2554	0.5135	−0.5389	−0.5124	−0.7017
0.1269	0.9827	−0.2652	−0.5645	0.3812	−0.3181	0.6370	−0.9764	−0.6817	−0.6304	0.9424	0.0069
−0.4123	0.0556	−0.8414	−0.4920	0.4873	0.3931	0.6202	−0.8650	0.3017	0.7456	0.0283	0.3789
−0.9717	−0.2941	−0.9094	−0.6815	−0.5724	0.9575	−0.9727	−0.4461	−0.1779	0.9563	−0.6917	0.8462
0.6046	−0.0979	−0.0292	−0.3385	0.6320	−0.3507	−0.3482	−0.1802	0.4422	0.8711	0.0372	0.1665

- Output Layer Weight Matrix:

$$\text{w\_output} = \begin{array}{rrrrrr} 0.9749 & 0.5933 & -0.7103 & 0.5541 & -0.6888 & -0.3538 \\ 0.0140 & 0.2826 & 0.9855 & 0.8707 & 0.4141 & 0.2090 \end{array}$$

Before any Changes

$$\text{w\_output} = \begin{array}{rrrrrr} 0.9749 & 0.5933 & -0.7103 & 0.5541 & -0.6888 & -0.3538 \\ 0.0140 & 0.2826 & 0.9855 & 0.8707 & 0.4141 & 0.2090 \end{array}$$

$$\text{z\_output} = 0.5047 \quad 1.501$$

$$\text{y\_output} = 1 \quad 1$$

Weight Modification at Output Layer:

- Neuron with Z closest to threshold

  z_index = 1

- Weights before change:

$$\text{w\_output\_min} = \begin{array}{rrrrrr} 0.9749 & 0.5933 & -0.7103 & 0.5541 & -0.6888 & -0.3538 \\ 0.0140 & 0.2826 & 0.9855 & 0.8707 & 0.4141 & 0.2090 \end{array}$$

- Weights after change:

$$\text{w\_output\_min} = \begin{array}{rrrrrr} 0.2549 & 1.3133 & 0.0097 & 1.2741 & -1.4088 & -0.3538 \\ 0.0140 & 0.2826 & 0.9855 & 0.8707 & 0.4141 & 0.2090 \end{array}$$

- Next Output Layer Neuron

  z_ind = 2

Final values for Output Layer after Convergence:

$$\text{w\_output} = \begin{array}{rrrrrr} 0.2549 & 1.3133 & 0.0097 & 1.2741 & -1.4088 & -0.3538 \\ -0.7060 & 1.0026 & 1.7055 & 1.5907 & -0.3059 & 0.2090 \end{array}$$

$$\text{z\_output} = 1.7970 \quad 3.0778$$

$$\text{y\_output} = 1 \quad 1$$

Final values for Hidden Layer after Convergence:

w_hidden =

Columns 1 through 12

−0.2630	1.3593	0.8750	0.4218	−0.0269	0.6532	0.8525	−1.6685	−0.5163	−0.5627	−0.5297	−1.5488
0.2329	−0.1504	0.6761	0.0423	0.6629	0.1875	0.1533	−0.1751	−0.6016	−0.9714	0.7511	−0.3460
0.9927	−0.4033	0.4272	0.8406	0.6071	0.5501	−0.3400	−0.8596	−0.7581	0.3686	−0.6020	−0.6334
0.8494	−0.7395	−0.2944	0.7219	−0.1397	−0.4833	0.5416	−0.8979	−0.1973	0.6348	−0.2891	0.5008
1.4906	1.6366	0.6425	0.3092	0.5427	0.0451	1.1972	−0.5929	−1.5854	0.0180	−0.7897	−0.2205
0.5930	−0.0853	0.8175	−0.0605	−0.7407	0.4429	0.6812	−0.7174	0.9599	−0.3352	−0.3762	−0.5934

Columns 13 through 24

1.2623	0.8311	1.4799	0.2762	0.2103	0.2320	0.5487	−1.4968	−0.8571	0.0047	−1.0030	−0.3003
−0.2570	−0.4116	−0.3409	0.5087	0.6436	−0.0342	−0.7515	−0.7608	0.2439	−0.8767	0.4824	−0.3426
0.6383	−0.0592	0.9073	0.0101	−0.2051	0.9051	−0.6792	0.4301	−0.7850	−0.1500	−0.2993	0.2404
−0.2520	0.2275	0.1467	0.3491	−0.5696	−0.7650	−0.3104	0.5042	−0.8040	0.5050	0.1335	0.1340
0.6257	1.6910	0.5158	0.1007	−0.1148	−0.3884	1.2018	−1.4992	−0.0983	0.2333	−0.3749	0.0545
−0.2432	−0.1404	−0.7061	−0.8046	−0.6752	0.6320	−0.2957	0.9080	0.5916	−0.7896	0.6390	0.4778

Columns 25 through 36

0.8916	−0.9563	0.1569	−0.0321	−0.1107	−1.0814	0.0596	−1.3715	−0.2802	−0.2583	−1.5253	−0.1338
0.7573	−0.4263	−0.6195	−0.4669	0.1387	−0.0657	−0.6288	−0.2554	0.5135	−0.5389	−0.5124	−0.7017
0.1269	0.9827	−0.2652	−0.5645	0.3812	−0.3181	0.6370	−0.9764	−0.6817	−0.6304	0.9424	0.0069
−0.4123	0.0556	−0.8414	−0.4920	0.4873	0.3931	0.6202	−0.8650	0.3017	0.7456	0.0283	0.3789
−0.2517	−1.0141	−1.6294	−1.4015	−1.2924	0.2375	−0.2527	−1.1661	−0.8979	0.2363	−1.4117	0.1262
0.6046	−0.0979	−0.0292	−0.3385	0.6320	−0.3507	−0.3482	−0.1802	0.4422	0.8711	0.0372	0.1665

z_hidden =	23.2709	6.8902	7.3169	0.6040	22.8362	−3.5097
y_hidden =	1	1	1	1	1	−1

Final Cumulative error

counter = 7

Training Efficiency

eff = 82.5000

Testing Procedure:

5 characters each for '0', 'C' and 'F' were used for testing the trained network. The network was found to detect 12 characters out of the 15 given characters resulting in an efficiency of 80%

Testing Efficiency:

eff = 80.0000%

5.A.5. *Conclusions and observations*

- The Neural Network was trained and tested for different test and training patterns. In all the cases the amount of convergence and error rate was observed.
- The convergence greatly depended on the hidden layers and number of neurons in each hidden layer.
- The number in each hidden layer should neither be too less or too high.
- The Neural network once properly trained was very accurate in classifying data in most of the test cases. The amount of error observed was 6%(approx.), which is ideal for classification problems like Face Detection.

5.A.6. MATLAB code for implementing MADALINE network:

Main Function:

```
% Training Patterns
X = train_pattern;
nu = 0.04;

% Displaying the 15 training patterns
figure(1)
for i = 1:15,
   subplot(5,3,i)
   display_image(X(:,i),6,6,1);
end

% Testing Patterns

Y = test_pattern;
nu = 0.04;

% Displaying the 15 testing patterns
figure(2)
for i = 1:15,
   subplot(5,3,i)
   display_image(Y(:,i),6,6,1);
end

% Initializations
index = zeros(2,6);
counter1 = 0;
counter2 = 0;

% Assign random weights initially at the start of training
w_hidden = (rand(6,36)-0.5)*2
w_output = (rand(2,6)-0.5)*2
%load w_hidden.mat
%load w_output.mat

% Function to calculate the parameters (z,y at the hidden and output
layers given the weights at the two layers)
[z_hidden, w_hidden, y_hidden, z_output, w_output, y_output, counter] =
calculation(w_hidden, w_output, X);

disp('Before Any Changes')

w_output
z_output
y_output

save z_output z_output;
save z_hidden z_hidden;
save y_hidden y_hidden;
save y_output y_output;

counter
```

```
%i = 1;
%min_z_output = min(abs(z_output));

disp('At counter minimum')

if (counter~= 0),
   [w_output_min,z_index] =
min_case(z_output,w_output,counter,y_hidden,nu);
   [z_hidden_min, w_hidden_min, y_hidden_min, z_output_min,
w_output_min, y_output_min, counter1] = calculation(w_hidden, w_output_min, X);
   counter1
end

w_output_min;
z_output_min;
y_output_min;

if (counter > counter1),
   %load w_output.mat;
   %load z_output.mat;
   %load y_output.mat;

   counter = counter1;
   w_output = w_output_min;
   z_output = z_output_min;
   y_output = y_output_min;
   index(2,z_index) = 1;
end

[w_output_max,z_ind] = max_case(z_output,w_output,counter,y_hidden,nu);
[z_hidden_max, w_hidden_max, y_hidden_max, z_output_max, w_output_max,
y_output_max, counter2] = calculation(w_hidden, w_output_max, X);

disp('At Counter minimum')

counter2
w_output_max;
z_output_max;
y_output_max;

if (counter2<counter),
   counter = counter2;
   w_output = w_output_max;
   z_output = z_output_max;
   y_output = y_output_max;
   index(2,z_ind) = 1;

end

% Adjusting the weights of the hidden layer
hidden_ind = zeros(1,6);
z_hid_asc = sort(abs(z_hidden));
for i = 1:6,
   for k = 1:6,
      if z_hid_asc(i) == abs(z_hidden(k)),
```

```
        hidden_ind(i) = k;
      end
   end
end

r1 = hidden_ind(1);
r2 = hidden_ind(2);
r3 = hidden_ind(3);
r4 = hidden_ind(4);
r5 = hidden_ind(5);
r6 = hidden_ind(6);

disp('At the beginning of the hidden layer Weight Changes - Neuron 1')

%load w_hidden.mat;

if ((counter~=0)&(counter>6)),
   [w_hidden_min] =
min_hidden_case(z_hidden,w_hidden,counter,X,nu,hidden_ind(1));
   [z_hidden_min, w_hidden_min, y_hidden_min, z_output_min, w_output,
y_output_min, counter3] = calculation(w_hidden_min, w_output, X);
   counter3
end

w_hidden;

if (counter3<counter),
   counter=counter3;
   w_hidden = w_hidden_min;
   y_hidden = y_hidden_min;
   z_hidden = z_hidden_min;
   z_output = z_output_min;
   y_output = y_output_min;
   index(1,r1) = 1;
end

disp('Hidden Layer - Neuron 2')
%load w_hidden.mat;
%counter=counter2;

if ((counter~=0)&(counter>6)),
   [w_hidden_min] =
min_hidden_case(z_hidden,w_hidden,counter,X,nu,hidden_ind(2));
   [z_hidden_min, w_hidden_min, y_hidden_min, z_output_min, w_output,
y_output_min, counter3] = calculation(w_hidden_min, w_output, X);
   counter3
end
w_hidden;
w_hidden_min;

if (counter3<counter),
   counter = counter3;
   w_hidden = w_hidden_min;
   y_hidden = y_hidden_min;
   z_hidden = z_hidden_min;
```

```
   z_output = z_output_min;
   y_output = y_output_min;
   index(1,r2)=1;
end

disp('Hidden Layer - Neuron 3')
%load w_hidden.mat;
%counter=counter2;

if ((counter~=0)&(counter>6)),
   [w_hidden_min] =
min_hidden_case(z_hidden,w_hidden,counter,X,nu,hidden_ind(3));
   [z_hidden_min, w_hidden_min, y_hidden_min, z_output_min, w_output,
y_output_min, counter3] = calculation(w_hidden_min, w_output, X);
   counter3
end

w_hidden;
w_hidden_min;

if (counter3<counter),
   counter = counter3;
   w_hidden = w_hidden_min;
   y_hidden = y_hidden_min;
   z_hidden = z_hidden_min;
   z_output = z_output_min;
   y_output = y_output_min;
   index(1,r3) = 1;
end

disp('Hidden Layer - Neuron 4')
%load w_hidden.mat;
%counter=counter2;

if ((counter~=0)&(counter>6)),
   [w_hidden_min] =
min_hidden_case(z_hidden,w_hidden,counter,X,nu,hidden_ind(4));
   [z_hidden_min, w_hidden_min, y_hidden_min, z_output_min, w_output,
y_output_min, counter3] = calculation(w_hidden_min, w_output, X);
   counter3
end

w_hidden;
w_hidden_min;

if (counter3<counter),
   counter = counter3;
   w_hidden = w_hidden_min;
   y_hidden = y_hidden_min;
   z_hidden = z_hidden_min;
   z_output = z_output_min;
   y_output = y_output_min;
   index(1,r4)=1;
end
disp('Hidden Layer - Neuron 5')
```

```
%load w_hidden.mat;
%counter=counter2;

if (counter~=0),
   [w_hidden_min] =
min_hidden_case(z_hidden,w_hidden,counter,X,nu,hidden_ind(5));
   [z_hidden_min, w_hidden_min, y_hidden_min, z_output_min, w_output,
y_output_min, counter3] = calculation(w_hidden_min, w_output, X);
   counter3
end
end
w_hidden;
w_hidden_min;

if (counter3<counter),
   counter = counter3;
   w_hidden = w_hidden_min;
   y_hidden = y_hidden_min;
   z_hidden = z_hidden_min;
   z_output = z_output_min;
   y_output = y_output_min;
   index(1,r5)=1;
end

disp('Combined Output Layer Neurons weight change');

%load w_hidden.mat;
%counter = counter2;
if ((counter~=0)&(index(2,[1:2])~=1)&(counter>6)),
   [w_output_two] =
min_output_double(z_hidden,y_hidden,counter,X,nu,w_output);
   [z_hidden_min, w_hidden_min, y_hidden_min, z_output_min, w_output,
y_output_min, counter3] = calculation(w_hidden,w_output_two, X);
   counter3
end
end
w_output;
%w_output_two;

if (counter3<counter),
   counter = counter3;
   %w_hidden = w_hidden_min;
   y_hidden = y_hidden_min;
   z_hidden = z_hidden_min;
   z_output = z_output_min;
   y_output = y_output_min;
   w_output = w_output_two;
end
disp('Begin 2 neuron changes - First Pair')

%load w_hidden.mat;
%counter = counter2;

if ((counter~=0)&(index(1,r1)~=1)&(index(1,r2)~=1)&(counter>6)),
   [w_hidden_two] =
```

```
min_hidden_double(z_hidden,w_hidden,counter,X,nu,hidden_ind(1),hidden_ind(2));
    [z_hidden_min, w_hidden_min, y_hidden_min, z_output_min, w_output,
y_output_min, counter3] = calculation(w_hidden_two, w_output, X);
    counter3
end
w_hidden;
w_hidden_min;

if (counter3<counter),
    counter = counter3;
    w_hidden = w_hidden_min;
    y_hidden = y_hidden_min;
    z_hidden = z_hidden_min;
    z_output = z_output_min;
    y_output = y_output_min;
end
disp('Begin 2 neuron changes - Second Pair')

%load w_hidden.mat;
%counter = counter2;

if ((counter~=0)&(index(1,r2)~=1)&(index(1,r3)~=1)&(counter>6)),
    [w_hidden_two] =
min_hidden_double(z_hidden,w_hidden,counter,X,nu,hidden_ind(2),hidden_ind(3));
    [z_hidden_min, w_hidden_min, y_hidden_min, z_output_min, w_output,
y_output_min, counter3] = calculation(w_hidden_two, w_output, X);
    counter3
end
w_hidden;
w_hidden_min;

if (counter3<counter),
    counter = counter3;
    w_hidden = w_hidden_min;
    y_hidden = y_hidden_min;
    z_hidden = z_hidden_min;
    z_output = z_output_min;
    y_output = y_output_min;
end
disp('Begin 2 neuron changes - Third Pair')

%load w_hidden.mat;
%counter = counter2;

if ((counter~=0)&(index(1,r3)~=1)&(index(1,r4)~=1)&(counter>6)),
    [w_hidden_two] =
min_hidden_double(z_hidden,w_hidden,counter,X,nu,hidden_ind(3),hidden_ind(4));
    [z_hidden_min, w_hidden_min, y_hidden_min, z_output_min, w_output,
y_output_min, counter3] = calculation(w_hidden_two, w_output, X);
    counter3
end

w_hidden;
w_hidden_min;
```

```
if (counter3<counter),
   counter = counter3;
   w_hidden = w_hidden_min;
   y_hidden = y_hidden_min;
   z_hidden = z_hidden_min;
   z_output = z_output_min;
   y_output = y_output_min;
end
disp('Begin 2 neuron changes - Fourth Pair')

%load w_hidden.mat;
%counter = counter2;

if ((counter~=0)&(index(1,r4)~=1)&(index(1,r5)~=1)&(counter>6)),
   [w_hidden_two] =
min_hidden_double(z_hidden,w_hidden,counter,X,nu,hidden_ind(4),hidden_ind(5));
   [z_hidden_min, w_hidden_min, y_hidden_min, z_output_min, w_output,
y_output_min, counter3] = calculation(w_hidden_two, w_output, X);
   counter3
end

w_hidden;
w_hidden_min;

disp('Final Values For Output')

w_output
z_output
y_output

disp('Final Values for Hidden')

w_hidden
z_hidden
y_hidden

disp('Final Error Number')

counter

disp('Efficiency')
eff = 100 - counter/40*100
```

Sub-functions:

```
*****************Function to calculate the parameters (z,y at the
hidden and output layers given the weights at the two layers)******************

function [z_hidden, w_hidden, y_hidden, z_output, w_output, y_output,
counter] = calculation(w_hidden, w_output, X)

% Outputs:
% z_hidden - hidden layer z value
```

```
% w_hidden - hidden layer weight
% y_hidden - hidden layer output
% Respecitvely for the output layers

% Inputs:
% Weights at the hidden and output layers and the training pattern set

counter = 0;

r = 1;
while(r<=15),

   r;
   for i = 1:6,
     z_hidden(i) = w_hidden(i,:)*X(:,r);

     if (z_hidden(i)>=0),
        y_hidden(i) = 1;
     else
        y_hidden(i) = -1;
     end %%End of If loop
   end %% End of for loop

   z_hidden;
   y_hiddent = y_hidden';

   for i = 1:2
      z_output(i) = w_output(i,:)*y_hiddent;
      if (z_output(i)>=0),
       y_output(i) = 1;
    else
       y_output(i) = -1;
     end %% End of If loop
     end%% End of for loop

     y_output;

   % Desired Output
   if (r<=5),
      d1 = [1 1]; % For 0
      else if (r>10),
            d1 = [-1 -1]  %For F
        else
      d1 = [-1 1]; % For C
      end
end
      for i = 1:2,
      error_val(i) = d1(i)-y_output(i);
      if (error_val(i)~=0),
         counter = counter+1;
      end
   end

r = r+1;
end
```

```
******Function to find weight changes for paired hidden layer**********

function [w_hidden_two] =
min_hidden_double(z_hidden,w_hidden,counter,X,nu,k,l)

w_hidden_two = w_hidden;
for j = 1:36,
   w_hidden_two(k,j) = w_hidden_two(k,j) + 2*nu*X(j,15)*counter;
   w_hidden_two(l,j) = w_hidden_two(l,j) + 2*nu*X(j,15)*counter;
end

*********Function to find weight changes at hidden layer**************

function [w_hidden_min] =
min_hidden_case(z_hidden,w_hidden,counter,X,nu,k)

w_hidden_min = w_hidden;
for j = 1:36,
   w_hidden_min(k,j) = w_hidden_min(k,j) + 2*nu*X(j,15)*counter;
end

%w_hidden_min


****Function to change weights for the max of 2z values at Output****

function [w_output_max,z_ind] =
max_case(z_output,w_output,counter,y_hidden,nu)

%load w_output;
%load z_output;
w_output_max = w_output;
z_ind = find(abs(z_output) == max(abs(z_output)))

for j = 1:5,
   w_output_max(z_ind,j) = w_output(z_ind,j)+2*nu*y_hidden(j)*counter;
   %  end
      %     z_output(z_index) = w_output(z_index,:)*y_hiddent;
end


****************Function to compute weight change at the output for
neuron whose Z value is close to the threshold**********************

function [w_output_min,z_index] =
min_case(z_output,w_output,counter,y_hidden,nu)

z_index = find(abs(z_output) == min(abs(z_output)))
w_output_min = w_output
for j = 1:5,
   w_output_min(z_index,j) = w_output(z_index,j) +
2*nu*y_hidden(j)*counter;
end

w_output_min
```

```
*******Function to find weight changes with paired output neurons******

function [w_output_two] =
min_output_double(z_hidden,y_hidden,counter,X,nu,w_output)

w_output_two = w_output;

for j = 1:6,
   w_output_two([1:2],j) = w_output([1:2],j)+2*nu*y_hidden(j)*counter;
end

y_hidden;
counter;
2*nu*y_hidden*counter;
```

Generating Training Set:

```
function X = train_pattern

 x1 = [1 1 1 1 1 1 ; 1 -1 -1 -1 -1 1; 1 -1 -1 -1 -1 1; 1 -1 -1 -1 -1 1;
      1 -1 -1 -1 -1 1; 1 1 1 1 1 1];
 x2 = [-1 1 1 1 1 1 ; 1 -1 -1 -1 -1 1; 1 -1 -1 -1 -1 1; 1 -1 -1 -1 -1 1;
      1 -1 -1 -1 -1 1; 1 1 1 1 1 1];
 x3 = [1 1 1 1 1 -1 ; 1 -1 -1 -1 -1 1; 1 -1 -1 -1 -1 1; 1 -1 -1 -1 -1 1;
      1 -1 -1 -1 -1 1; 1 1 1 1 1 1];
 x4 = [1 1 1 1 1 1 ; 1 -1 -1 -1 -1 1; 1 -1 -1 -1 -1 1; 1 -1 -1 -1 -1 1;
      1 -1 -1 -1 -1 1; -1 1 1 1 1 1];
 x5 = [-1 1 1 1 1 -1 ; 1 -1 -1 -1 -1 1; 1 -1 -1 -1 -1 1; 1 -1 -1 -1 -1 1;
      1 -1 -1 -1 -1 1; 1 1 1 1 1 1];
 x6 = [1 1 1 1 1 1 ; 1 -1 -1 -1 -1 -1; 1 -1 -1 -1 -1 -1; 1 -1 -1 -1 -1 -1;
      1 -1 -1 -1 -1 -1; 1 1 1 1 1 1];
 x7 = [-1 1 1 1 1 1 ; 1 -1 -1 -1 -1 -1; 1 -1 -1 -1 -1 -1; 1 -1 -1 -1 -1 -1;
      1 -1 -1 -1 -1 -1; 1 1 1 1 1 1];
 x8 = [1 1 1 1 1 1 ; 1 -1 -1 -1 -1 -1; 1 -1 -1 -1 -1 -1; 1 -1 -1 -1 -1 -1;
      1 -1 -1 -1 -1 -1; -1 1 1 1 1 1];
 x9 = [1 1 1 1 1 -1 ; 1 -1 -1 -1 -1 -1; 1 -1 -1 -1 -1 -1; 1 -1 -1 -1 -1 -1;
      1 -1 -1 -1 -1 -1;1 1 1 1 1 -1];
x10 = [-1 1 1 1 1 1 ; 1 -1 -1 -1 -1 -1; 1 -1 -1 -1 -1 -1; 1 -1 -1 -1 -1 -1;
       1 -1 -1 -1 -1 -1; 1 1 1 1 1 -1];
x11 = [1 1 1 1 1 1 ; 1 -1 -1 -1 -1 -1; 1 1 1 1 1 1; 1 -1 -1 -1 -1 -1;
       1 -1 -1 -1 -1 -1; 1 -1 -1 -1 -1 -1];
x12 = [-1 1 1 1 1 1 ; 1 -1 -1 -1 -1 -1; 1 1 1 1 1 1; 1 -1 -1 -1 -1 -1;
       1 -1 -1 -1 -1 -1; 1 -1 -1 -1 -1 -1];
x13 = [1 1 1 1 1 -1 ; 1 -1 -1 -1 -1 -1; 1 1 1 1 1 -1; 1 -1 -1 -1 -1 -1;
       1 -1 -1 -1 -1 -1; 1 -1 -1 -1 -1 -1];
x14 = [1 1 1 1 1 1 ; 1 -1 -1 -1 -1 -1; -1 1 1 1 1 1; 1 -1 -1 -1 -1 -1;
       1 -1 -1 -1 -1 -1; 1 -1 -1 -1 -1 -1];
x15 = [1 1 1 1 1 1 ; 1 -1 -1 -1 -1 -1; 1 1 1 1 1 -1; 1 -1 -1 -1 -1 -1;
       1 -1 -1 -1 -1 -1; 1 -1 -1 -1 -1 -1];

xr1 = reshape(x1',1,36);
xr2 = reshape(x2',1,36);
xr3 = reshape(x3',1,36);
xr4 = reshape(x4',1,36);
```

```
 xr5 = reshape(x5',1,36);
 xr6 = reshape(x6',1,36);
 xr7 = reshape(x7',1,36);
 xr8 = reshape(x8',1,36);
 xr9 = reshape(x9',1,36);
xr10 = reshape(x10',1,36);
xr11 = reshape(x11',1,36);
xr12 = reshape(x12',1,36);
xr13 = reshape(x13',1,36);
xr14 = reshape(x14',1,36);
xr15 = reshape(x15',1,36);

X = [xr1' xr2' xr3' xr4' xr5' xr6' xr7' xr8' xr9' xr10' xr11' xr12' xr13'
     xr14' xr15'];
```

Generating Test Set:

```
function [X_test] = test_pattern

 X1 = [1 1 1 -1 1 1 ; 1 -1 -1 -1 -1 1; 1 -1 -1 -1 -1 1; 1 -1 -1 -1 -1 -1;
       1 -1 -1 -1 -1 1; 1 1 1 -1 1 1];
 X2 = [1 1 1 1 1 -1 ; 1 -1 -1 -1 -1 1; 1 -1 -1 -1 -1 1; 1 -1 -1 -1 -1 1;
       1 -1 -1 -1 -1 1; -1 1 1 1 1 1];
 X3 = [-1 1 1 1 1 1 ; 1 -1 -1 -1 -1 1; 1 -1 -1 -1 -1 1; 1 -1 -1 -1 -1 1;
       1 -1 -1 -1 -1 1; -1 1 1 1 1 1];
 X4 = [1 1 1 1 1 1 ; 1 -1 -1 -1 -1 1; 1 -1 -1 -1 -1 1; 1 -1 -1 -1 -1 1;
       -1 1 -1 -1 -1 1; -1 -1 1 1 1 -1];
 X5 = [-1 1 1 1 -1 -1 ; 1 -1 -1 -1 1 1; 1 -1 -1 -1 -1 1; 1 -1 -1 -1 -1 1;
       1 -1 -1 -1 -1 1; -1 1 1 1 1 1];
 X6 = [-1 -1 1 1 1 1 ; -1 1 -1 -1 -1 -1; 1 -1 -1 -1 -1 -1; 1 -1 -1 -1 -1 -1;
       -1 1 -1 -1 -1 -1; -1 -1 1 1 1 1];
 X7 = [1 1 1 1 -1 -1 ; 1 -1 -1 -1 -1 -1; 1 -1 -1 -1 -1 -1; 1 -1 -1 -1 -1 -1;
       1 -1 -1 -1 -1 -1; 1 1 1 1 1 1];
 X8 = [1 1 1 1 1 1 ; 1 -1 -1 -1 -1 -1; 1 -1 -1 -1 -1 -1; 1 -1 -1 -1 -1 -1;
       1 -1 -1 -1 -1 -1; 1 1 1 1 -1 -1];
 X9 = [1 1 1 1 1 -1 ; 1 -1 -1 -1 -1 -1; 1 -1 -1 -1 -1 -1; 1 -1 -1 -1 -1 -1;
       -1 1 -1 -1 -1 -1;-1 -1 1 1 1 -1];
X10 = [-1 1 1 1 1 1 ; 1 -1 -1 -1 -1 -1; 1 -1 -1 -1 -1 -1; 1 -1 -1 -1 -1 -1;
       1 -1 -1 -1 -1 -1; -1 -1 1 1 1 -1];
X11 = [-1 1 1 1 1 1 ; 1 -1 -1 -1 -1 -1; 1 1 1 -1 -1 -1; 1 -1 -1 -1 -1 -1;
       1 -1 -1 -1 -1 -1; 1 -1 -1 -1 -1 -1];
X12 = [1 1 1 1 1 1 ; -1 -1 -1 -1 -1 -1; 1 1 1 1 1 1; 1 -1 -1 -1 -1 -1;
       1 -1 -1 -1 -1 -1; 1 -1 -1 -1 -1 -1];
X13 = [1 1 1 -1 -1 -1 ; 1 -1 -1 -1 -1 -1; 1 1 1 1 1 -1; 1 -1 -1 -1 -1 -1;
       1 -1 -1 -1 -1 -1; 1 -1 -1 -1 -1 -1];
X14 = [1 1 -1 1 1 -1 ; 1 -1 -1 -1 -1 -1; -1 -1 1 1 1 1; 1 -1 -1 -1 -1 -1;
       1 -1 -1 -1 -1 -1; -1 -1 -1 -1 -1 -1];
X15 = [-1 -1 1 1 1 1 ; -1 1 -1 -1 -1 -1; -1 1 1 1 1 -1; 1 -1 -1 -1 -1 -1;
       1 -1 -1 -1 -1 -1; 1 -1 -1 -1 -1 -1];

 xr1 = reshape(X1',1,36);
 xr2 = reshape(X2',1,36);
 xr3 = reshape(X3',1,36);
 xr4 = reshape(X4',1,36);
```

```
 xr5 = reshape(X5',1,36);
 xr6 = reshape(X6',1,36);
 xr7 = reshape(X7',1,36);
 xr8 = reshape(X8',1,36);
 xr9 = reshape(X9',1,36);
xr10 = reshape(X10',1,36);
xr11 = reshape(X11',1,36);
xr12 = reshape(X12',1,36);
xr13 = reshape(X13',1,36);
xr14 = reshape(X14',1,36);
xr15 = reshape(X15',1,36);


X_test = [xr1' xr2' xr3' xr4' xr5' xr6' xr7' xr8' xr9' xr10' xr11' xr12' xr13'
          xr14' xr15'];
```

Chapter 6

Back Propagation

6.1. The Back Propagation Learning Procedure

The back propagation (BP) algorithm was proposed in 1986 by Rumelhart, Hinton and Williams for setting weights and hence for the training of multi-layer perceptrons. This opened the way for using multi-layer ANNs, nothing that the hidden layers have no desired (hidden) outputs accessible. Once the BP algorithm of Rumelhart *et al.* was published, it was very close to algorithms proposed earlier by Werbos in his Ph.D. dissertation in Harvard in 1974 and then in a report by D. B. Parker at Stanford in 1982, both unpublished and thus unavailable to the community at large. It goes without saying that the availability of a rigorous method to set intermediate weights, namely to train hidden layers of ANNs gave a major boost to the further development of ANN, opening the way to overcome the single-layer shortcomings that had been pointed out by Minsky and which nearly dealt a death blow to ANNs.

6.2. Derivation of the BP Algorithm

The BP algorithm starts, of necessity with computing the *output layer*, which is the only one where desired outputs *are* available, but the outputs of the intermediate layers are unavailable (see Fig. 6.1), as follows:

Let ε denote the error-energy at the output layer, where:

$$\varepsilon \triangleq \frac{1}{2}\sum_k (d_k - y_k)^2 = \frac{1}{2}\sum_k e_k^2 \tag{6.1}$$

$k = 1 \cdots N$; N being the number of neurons in the output layer. Consequently, a gradient of ε is considered, where:

$$\nabla\varepsilon_k = \frac{\partial\varepsilon}{\partial w_{kj}} \tag{6.2}$$

Now, by the steepest descent (gradient) procedure, as in Sec. 3.4.2, we have that

$$w_{kj}(m+1) = w_{kj}(m) + \Delta w_{kj}(m) \tag{6.3}$$

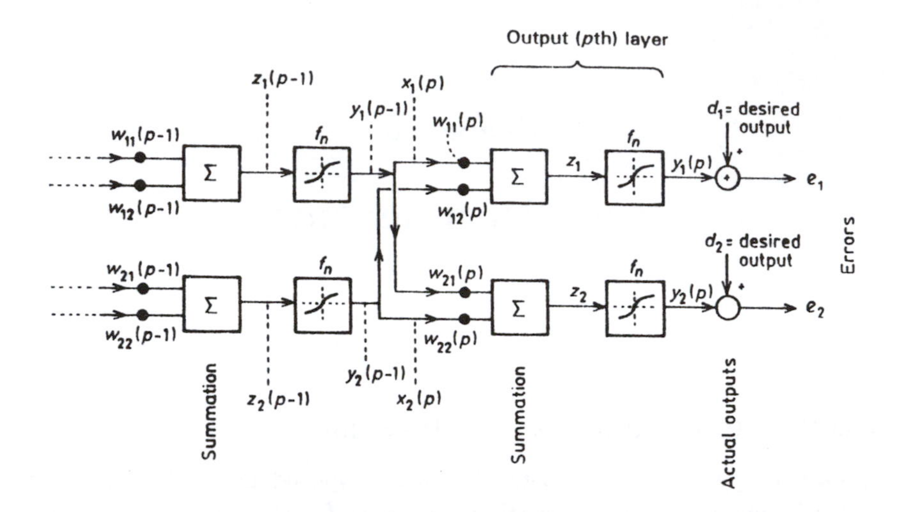

Fig. 6.1. A multi-layer perceptron.

j denoting the jth input to the kth neuron of the output layer, where, again by the steepest descent procedure:

$$\Delta w_{kj} = -\eta \frac{\partial \varepsilon}{\partial w_{kj}} \tag{6.4}$$

The minus (−) sign in Eq. (6.4) indicates a down-hill direction towards a minimum. We note from the perceptron's definition that the k's perceptron's node-output z_k is given by

$$z_k = \sum_j w_{kj} x_j \tag{6.5}$$

x_j being the jth input to that neuron, and noting that the perceptron's output y_k is:

$$y_k = F_N(z_k) \tag{6.6}$$

F being a nonlinear function as discussed in Chap. 5. We now substitute

$$\frac{\partial \varepsilon}{\partial w_{kj}} = \frac{\partial \varepsilon}{\partial z_k} \frac{\partial z_k}{\partial w_{kj}} \tag{6.7}$$

and, by Eq. (6.5):

$$\frac{\partial z_k}{\partial w_{kj}} = x_j(p) = y_j(p-1) \tag{6.8}$$

p denoting the output layer, such that Eq. (6.7) becomes:

$$\frac{\partial \varepsilon}{\partial w_{kj}} = \frac{\partial \varepsilon}{\partial z_k} x_j(p) = \frac{\partial \varepsilon}{\partial z_r} y_j(p-1) \tag{6.9}$$

Defining:

$$\Phi_k(p) = -\frac{\partial \varepsilon}{\partial z_k(p)} \tag{6.10}$$

then Eq. (6.9) yields:

$$\frac{\partial \varepsilon}{\partial w_{kj}} = -\Phi_k(p) x_j(p) = -\Phi_k y_j(p-1) \tag{6.11}$$

and, by Eqs. (6.4) and (6.11):

$$\Delta w_{kj} = \eta \phi_k(p) x_j(p) = \eta \Phi_k(p) y_j(p-1) \tag{6.12}$$

j denoting the jth input to neuron k of the output (p) layer.

Furthermore, by Eq. (6.10):

$$\Phi_k = -\frac{\partial \varepsilon}{\partial z_k} = -\frac{\partial \varepsilon}{\partial y_k} \frac{\partial y_k}{\partial z_k} \tag{6.13}$$

But, by Eq. (6.1):

$$\frac{\partial \varepsilon}{\partial y_k} = -(d_k - y_k) = y_k - d_k \tag{6.14}$$

whereas, for a sigmoid nonlinearity:

$$y_k = F_N(z_k) = \frac{1}{1 + \exp(-z_k)} \tag{6.15}$$

we have that:

$$\frac{\partial y_k}{\partial z_k} = y_k(1 - y_k) \tag{6.16}$$

Consequently; by Eqs. (6.13), (6.14) and (6.16):

$$\Phi_k = y_k(1 - y_k)(d_k - y_k) \tag{6.17}$$

such that, at the output layer, by Eqs. (6.4), (6.7):

$$\Delta w_{kj} = -\eta \frac{\partial \varepsilon}{\partial w_{kj}} = -\eta \frac{\partial \varepsilon}{\partial z_k} \frac{\partial z_k}{\partial w_{kj}} \tag{6.18}$$

where, by Eqs. (6.8) and (6.13)

$$\Delta w_{kj}(p) = \eta \Phi_k(p) y_j(p-1) \tag{6.19}$$

Φ_k being as in Eq. (6.17), to complete the derivation of the setting of output layer weights.

Back-propagating to the rth hidden layer, we still have, as before

$$\Delta w_{ji} = -\eta \frac{\partial \varepsilon}{\partial w_{ji}} \tag{6.20}$$

for the ith branch into the jth neuron of the rth hidden layer. Consequently, in parallelity to Eq. (6.7):

$$\Delta w_{ji} = -\eta \frac{\partial \varepsilon}{\partial z_j} \frac{\partial z_j}{\partial w_{ji}} \tag{6.21}$$

and noting Eq. (6.8) and the definition of Φ in Eq. (6.13):

$$\Delta w_{ji} = -\eta \frac{\partial \varepsilon}{\partial z_j} y_i(r-1) = \eta \Phi_j(r) y_i(r-1) \tag{6.22}$$

such that, by the right hand-side relation of Eq. (6.13)

$$\Delta w_{ji} = -\eta \left[\frac{\partial \varepsilon}{\partial y_j(r)} \frac{\partial y_j}{\partial z_j} \right] y_i(r-1) \tag{6.23}$$

where $\frac{\partial \varepsilon}{\partial y_j}$ is *inaccessible* (as is, therefore, also $\Phi_j(r)$ above).

However, ε can only be affected by upstream neurons when one propagates *backwards* from the output. No other information is available at that stage. Therefore:

$$\frac{\partial \varepsilon}{\partial y_j(r)} = \sum_k \frac{\partial \varepsilon}{\partial z_k(r+1)} \left[\frac{\partial z_k(r+1)}{\partial y_j(r)} \right] = \sum_k \frac{\partial \varepsilon}{\partial z_k} \left[\frac{\partial}{\partial y_j(r)} \sum_m w_{km}(r+1) y_m(r) \right] \tag{6.24}$$

where the summation over k is performed over the neurons of the next (the $r+1$) layer that connect to $y_j(r)$, whereas summation over m is over all inputs to each k'th neuron of the $(r+1)$ layer.

Hence, and noting the definition of Φ, Eq. (6.24) yields:

$$\frac{\partial \varepsilon}{\partial y_j(r)} = \sum_k \frac{\partial \varepsilon}{\partial z_k(r+1)} w_{kj} = -\sum_k \Phi_k(r+1) w_{kj}(r+1) \tag{6.25}$$

since only $w_{k_j}(r+1)$ is connected to $y_j(r)$.

Consequently, by Eqs. (6.13), (6.16) and (6.25):

$$\begin{aligned} \Phi_j(r) &= \frac{\partial y_j}{\partial z_j} \sum_k \Phi_k(r+1) w_{kj}(r+1) \\ &= y_j(r)[1 - y_j(r)] \sum_k \Phi_k(r+1) w_{kj}(r+1) \end{aligned} \tag{6.26}$$

and, via Eq. (6.19):

$$\Delta w_{ji}(r) = \eta \Phi_j(r) y_i(r-1) \tag{6.27}$$

to obtain $\Delta w_{ji}(r)$ as a function of ϕ and the weights of the $(r+1)$ layer, noting Eq. (6.26).

Note that we cannot take partial derivatives of ε with respect to the hidden layer considered. We thus must take the partial derivatives of ε with respect to the variables upstream in the direction of the output, which are the only ones that affect ε. This observation is the basis for the Back-Propagation procedure, to facilitate overcoming the lack of accessible error data in the hidden layers.

The BP algorithm thus propagates backwards all the way to $r = 1$ (the first layer), to complete its derivation. Its computation can thus be summarized as follows:

Apply the first training vector. Subsequently, compute $\Delta w_{kj}(p)$ from Eqs. (6.17) and (6.19) for the output (the p) layer and then proceed through computing $\Delta w_{ji}(r)$ from Eq. (6.27) for $r = p-1, p-2, \ldots, 2, 1$; using Eq. (6.26) to update $\Phi_j(r)$ on the basis of $\Phi_j(r+1)$ upstream (namely back-propagating from layer $r+1$ to layer r), etc. Next, update $w(m+1)$ from $w(m)$ and $\Delta w(m)$ for the $m+1$ iteration via Eq. (6.3) for the latter training set. Repeat the whole process when applying the next training vector until you go through all L training vectors, then repeat the whole process for the next $(m+2)$, $(m+3)\ldots.$ iteration until adequate convergence is reached.

The *learning rate* η should be adjusted stepwise, considering out comment at the end of Sec. 3.4.2. However, since convergence is considerably faster than in Adaline/Madaline designs, when the error becomes very small, it is advisable to reinstate η to its initial value before proceeding.

Initialization of $w_{ji}(o)$ is accomplished by setting each weight to a low-valued random value selected from a pool of random numbers, say in the range from -5 to $+5$.

As in the case of the Madaline network of Sec. 5, the number of hidden layer neurons should be higher rather than lower. However, for simple problems, one or two hidden layers may suffice.

6.3. Modified BP Algorithms

6.3.1. *Introduction of bias into NN*

It is often advantageous to apply some bias to the neurons of a neural network (see Fig. 6.2). The bias can be trainable when associated with a trainable weight to be modified as is any other weight. Hence the bias is realized in terms of an input with some constant (say $+1$ or $+B$) input, and the exact bias b_i (at the ith neuron) is then given

$$b_i = w_{oi}B \tag{6.28}$$

w_{oi} being the weight of the bias term at the input to neuron i (see Fig. 7). Note that the bias may be positive or negative, depending on its weight.

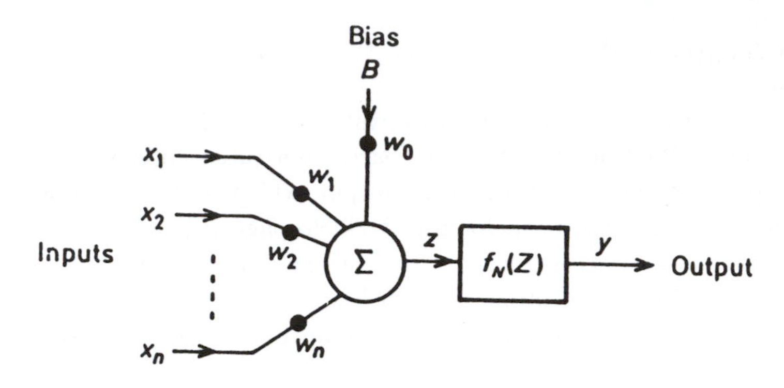

Fig. 6.2. A biased neuron.

6.3.2. *Incorporating momentum or smoothing to weight adjustment*

The backpropagation (BP) algorithm to compute weights of neurons may tend to instability under certain operation conditions. To reduce the tendency to instability Rumelhart *et al.* (1986) suggested to add a momentum term to Eq. (6.1). Hence, Eq. (6.12) is modified to:

$$\Delta w_{ij}^{(m)} = \eta \Phi_i(r) y_j(r-1) + \alpha \Delta w_{ij}^{(m-1)} \tag{6.29}$$

$$w_{ij}^{(m+1)} = w_{ij}^{(m)} + \Delta w_{ij}^{(m)} \tag{6.30}$$

for the $m+1$ iteration, with $0 < \alpha < 1$; α being the momentum coefficient (usually around 0.9). The employment of α will tend to avoid fast fluctuations, but it may not always work, or could even harm convergence.

Another smoothing method, for the same purpose and also not always advisable, is that of employing a smoothing term as proposed by Sejnowski and Rosenberg (1987), is given as follows:

$$\Delta w_{ij}^{(m)} = \alpha \Delta w_{ij}^{(m-1)} + (1-\alpha) \Phi_i(r) y_j(r-1) \tag{6.31}$$

$$w_{ij}^{(m+1)} = w_{ij}^{(m)} + \eta \Delta w_{ij}^{(m)} \tag{6.32}$$

with $0 < \alpha < 1$. Note that for $\alpha = 0$ no smoothing takes place whereas causes the algorithm to get stuck. η of Eq. (6.32) is again between 0 and 1.

6.3.3. *Other modification concerning convergence*

Improved convergence of the BP algorithm can often be achieved by *modifying the range of the sigmoid* function from the range of zero to one, to a range from -0.5 to $+0.5$. Introduction of feedback (see Ch. 12) may sometimes be used.

Modifying step size can be employed to avoid the BP algorithm from *getting stuck* (learning paralysis) or from oscillating. This is often achieved by reducing step size, at least when the algorithm approached paralysis or when it starts oscillating.

Convergence to *local minima* can best be avoided by statistical methods where there always exists a finite probability of moving the network away from an apparent or a real minimum by a large step.

Modified (resilient) BP algorithms, such as RPROP (Riedmiller and Braun, 1993) greatly speed up convergence and reduce sensitivity to initialization. It considers only signs of partial derivatives to compute weights by BP, rather than their values.

6.A. Back Propagation Case Study*: Character Recognition

6.A.1. *Introduction*

We are trying to solve a simple character recognition problem using a network of perceptrons with back propagation learning procedure. Our task is to teach the neural network to recognize 3 characters, that is, to map them to respective pairs {0,1}, {1,0} and {1,1}. We would also like the network to produce a special error signal {0,0} in response to any other character.

6.A.2. *Network design*

(a) Structure: The neural network of the present design consists of three layers with 2 neurons each, one output layer and two hidden layers. There are 36 inputs to the network. In this particular case the sigmoid function:

$$y = \frac{1}{1 + \exp(-z)}$$

is chosen as a nonlinear neuron activation function. Bias terms (equal to 1) with trainable weights were also included in the network structure. The structural diagram of the neural network is given in Fig. 6.A.1.

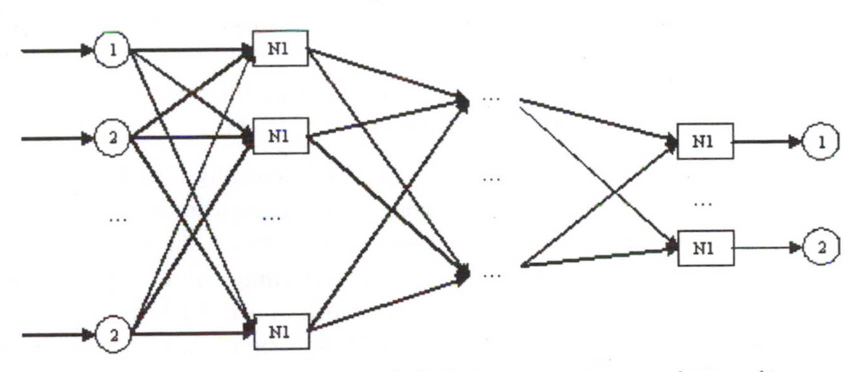

Fig. 6.A.1. Schematic design of the back-propagation neural network.

*Computed by Maxim Kolesnikov, ECE Dept., University of Illinois, Chicago, 2005.

(b) Dataset Design: We teach the neural network to recognize characters 'A', 'B' and 'C'. To train the network to produce error signal we will use another 6 characters 'D', 'E', 'F', 'G', 'H' and 'I'. To check whether the network has learned to recognize errors we will use characters 'X', 'Y' and 'Z'. Note that we are interested in checking the response of the network to errors on the characters which were not involved in the training procedure.

The characters to be recognized are given on a 6×6 grid. Each of the 36 pixels is set to either 0 or 1. The Corresponding 6×6 matrices of the character representation is given as:

A:	001100	B:	111110	C:	011111
	010010		100001		100000
	100001		111110		100000
	111111		100001		100000
	100001		100001		100000
	100001		111110		011111
D:	111110	E:	111111	F:	111111
	100001		100000		100000
	100001		111111		111111
	100001		100000		100000
	100001		100000		100000
	111110		111111		100000
G:	011111	H:	100001	I:	001110
	100000		100001		000100
	100000		111111		000100
	101111		100001		000100
	100001		100001		000100
	011111		100001		001110
X:	100001	Y:	010001	Z:	111111
	010010		001010		000010
	001100		000100		000100
	001100		000100		001000
	010010		000100		010000
	100001		000100		111111

(c) Network Set-Up: The Back propagation (BP) learning algorithm of Section 6.2 was used to solve the problem. The goal of this algorithm is to minimize the error-energy at the output layer, as in Sect. 6.2 above, using equations (6.17), (6.19), (6.26), (6.27) thereof. In this method a training set of input vectors is applied vector-by-vector to the input of the network and is forward-propagated to the output. Weights are then adjusted by the BP algorithm as above. Subsequently, we repeat these steps for all training sets. The whole process is then repeated for the next $(m + 2)$-th iteration and so on. We stop when adequate convergence is reached.

The program code in C++ was written to simulate the response of the network and perform the learning procedure, as in Section 6.A.5 below.

6.A.3. *Results*

(a) Network Training: To train the network to recognize the above characters we applied the corresponding 6×6 grids in the form of 1×36 vectors to the input of the network. The character was considered recognized if both outputs of the network were no more than 0.1 off their respective desired values. The initial learning rate η was experimentally set to 1.5 and was decreased by a factor of 2 after each 100th iteration. This approach, however, resulted in the learning procedure getting stuck in various local minima. We tried running the learning algorithm for 1000 iterations and it became clear that the error-energy parameter had converged to some steady value, but recognition failed for all characters (vectors).

However, none of our training vectors were recognized at this point:

TRAINING VECTOR 0: [0.42169 0.798603] — NOT RECOGNIZED —
TRAINING VECTOR 1: [0.158372 0.0697667] — NOT RECOGNIZED —
TRAINING VECTOR 2: [0.441823 0.833824] — NOT RECOGNIZED —
TRAINING VECTOR 3: [0.161472 0.0741904] — NOT RECOGNIZED —
TRAINING VECTOR 4: [0.163374 0.0769596] — NOT RECOGNIZED —
TRAINING VECTOR 5: [0.161593 0.074359] — NOT RECOGNIZED —
TRAINING VECTOR 6: [0.172719 0.0918946] — NOT RECOGNIZED —
TRAINING VECTOR 7: [0.15857 0.0700591] — NOT RECOGNIZED —
TRAINING VECTOR 8: [0.159657 0.0719576] — NOT RECOGNIZED —

Training vectors $0, 1, \ldots, 8$ in these log entries correspond to the characters 'A', 'B', ..., 'I'.

To prevent this from happening, one more modification was made. After each 400th iteration we reset the learning rate to its initial value. Then after about 2000 iterations we were able to converge to 0 error and to correctly recognize all characters:

TRAINING VECTOR 0: [0.0551348 0.966846] — RECOGNIZED —
TRAINING VECTOR 1: [0.929722 0.0401743] — RECOGNIZED —
TRAINING VECTOR 2: [0.972215 0.994715] — RECOGNIZED —
TRAINING VECTOR 3: [0.0172118 0.00638034] — RECOGNIZED —
TRAINING VECTOR 4: [0.0193525 0.00616272] — RECOGNIZED —
TRAINING VECTOR 5: [0.00878156 0.00799531] — RECOGNIZED —
TRAINING VECTOR 6: [0.0173236 0.00651032] — RECOGNIZED —
TRAINING VECTOR 7: [0.00861903 0.00801831] — RECOGNIZED —
TRAINING VECTOR 8: [0.0132965 0.00701945] — RECOGNIZED —

(b) Recognition Results: In order to determine if error detection is performed correctly, we saved the obtained weights into a data file, modified the datasets in the program replacing the characters 'G', 'H' and 'I' (training vectors 6, 7 and 8) by the characters 'X', 'Y' and 'Z'. We then ran the program, loaded the previously saved weights from the data file and applied the input to the network. Note that we performed no further training. We got the following results:

TRAINING VECTOR 6: [0.00790376 0.00843078] — RECOGNIZED —
TRAINING VECTOR 7: [0.0105325 0.00890258] — RECOGNIZED —
TRAINING VECTOR 8: [0.0126299 0.00761764] — RECOGNIZED —

All three characters were successfully mapped to error signal 0,0.

(c) Robustness Investigation: To investigate how robust our neural network was, we added some noise to the input and got the following results. In the case of 1-bit distortion (out of 36 bits) the recognition rates were:

TRAINING VECTOR 0: 25/36 recognitions (69.4444%)
TRAINING VECTOR 1: 33/36 recognitions (91.6667%)
TRAINING VECTOR 2: 32/36 recognitions (88.8889%)
TRAINING VECTOR 3: 35/36 recognitions (97.2222%)
TRAINING VECTOR 4: 34/36 recognitions (94.4444%)
TRAINING VECTOR 5: 35/36 recognitions (97.2222%)
TRAINING VECTOR 6: 36/36 recognitions (100%)
TRAINING VECTOR 7: 35/36 recognitions (97.2222%)
TRAINING VECTOR 8: 36/36 recognitions (100%)

We also investigated the case of 2-bit distortion and were able to achieve the following recognition rates:

TRAINING VECTOR 0: 668/1260 recognitions (53.0159%)
TRAINING VECTOR 1: 788/1260 recognitions (62.5397%)
TRAINING VECTOR 2: 906/1260 recognitions (71.9048%)
TRAINING VECTOR 3: 1170/1260 recognitions (92.8571%)
TRAINING VECTOR 4: 1158/1260 recognitions (91.9048%)
TRAINING VECTOR 5: 1220/1260 recognitions (96.8254%)
TRAINING VECTOR 6: 1260/1260 recognitions (100%)
TRAINING VECTOR 7: 1170/1260 recognitions (92.8571%)
TRAINING VECTOR 8: 1204/1260 recognitions (95.5556%)

6.A.4. *Discussion and conclusions*

We were able to train our neural network so that it successfully eecognizes the three given characters and at the same time is able to classify other characters as errors. However, there is a price to pay for this convenience. It seems that the greater the error detection rate is, the less robust our network is. For instance, when 2 bits of character 'A' are distorted, the network has only 53% recognition rate. Roughly speaking, in 1 out of 2 such cases, the network 'thinks' that its input is not the symbol 'A' and therefore must be classified as error. Overall, the back propagation network proved to be much more powerful than Madaline. It is possible to achieve convergence much faster and it is also easier to program. There are cases, however, when the back propagation learning algorithm gets stuck in a local minimum but they can be successfully dealt with by tuning the learning rate and the law of changing learning rate during the learning process for each particular problem.

6.A.5. *Program Code (C++)*

```
/*
*/
#include<math.h>
#include<iostream>
#include<fstream>
using namespace std;
#define N_DATASETS 9
#define N_INPUTS 36
#define N_OUTPUTS 2
#define N_LAYERS 3
// {# inputs, # of neurons in L1, # of neurons in L2, # of neurons in L3}
short conf[4] = {N_INPUTS, 2, 2, N_OUTPUTS};
float **w[3], *z[3], *y[3], *Fi[3], eta;     // According to the number of
layers ofstream ErrorFile("error.txt",  ios::out);
// 3 training sets
bool dataset[N_DATASETS][N_INPUTS] = {
{ 0, 0, 1, 1, 0, 0,          //  'A'
  0, 1, 0, 0, 1, 0,
  1, 0, 0, 0, 0, 1,
  1, 1, 1, 1, 1, 1,
  1, 0, 0, 0, 0, 1,
  1, 0, 0, 0, 0, 1},
{ 1, 1, 1, 1, 1, 0,          //  'B'
  1, 0, 0, 0, 0, 1,
  1, 1, 1, 1, 1, 0,
  1, 0, 0, 0, 0, 1,


  1, 0, 0, 0, 0, 1,
  1, 1, 1, 1, 1, 0},
{ 0, 1, 1, 1, 1, 1,          //  'C'
  1, 0, 0, 0, 0, 0,
  1, 0, 0, 0, 0, 0,
  1, 0, 0, 0, 0, 0,
  1, 0, 0, 0, 0, 0,
  0, 1, 1, 1, 1, 1},
{ 1, 1, 1, 1, 1, 0,          //  'D'
  1, 0, 0, 0, 0, 1,
  1, 0, 0, 0, 0, 1,
  1, 0, 0, 0, 0, 1,
```

```
  1,  0,  0,  0,  0,  1,
  1,  1,  1,  1,  1,  0},
{ 1,  1,  1,  1,  1,  1,          //  'E'
  1,  0,  0,  0,  0,  0,
  1,  1,  1,  1,  1,  1,
  1,  0,  0,  0,  0,  0,
  1,  0,  0,  0,  0,  0,
  1,  1,  1,  1,  1,  1},
{ 1,  1,  1,  1,  1,  1,          //  'F'
  1,  0,  0,  0,  0,  0,
  1,  1,  1,  1,  1,  1,
  1,  0,  0,  0,  0,  0,
  1,  0,  0,  0,  0,  0,
  1,  0,  0,  0,  0,  0},
{ 0,  1,  1,  1,  1,  1,          //  'G'
  1,  0,  0,  0,  0,  0,
  1,  0,  0,  0,  0,  0,
  1,  0,  1,  1,  1,  1,
  1,  0,  0,  0,  0,  1,
  0,  1,  1,  1,  1,  1},
{ 1,  0,  0,  0,  0,  1,          //  'H'
  1,  0,  0,  0,  0,  1,
  1,  1,  1,  1,  1,  1,
  1,  0,  0,  0,  0,  1,
  1,  0,  0,  0,  0,  1,
  1,  0,  0,  0,  0,  1},
{ 0,  0,  1,  1,  1,  0,          //  'I'
  0,  0,  0,  1,  0,  0,
  0,  0,  0,  1,  0,  0,
  0,  0,  0,  1,  0,  0,
  0,  0,  0,  1,  0,  0,
  0,  0,  1,  1,  1,  0}

// Below are the datasets for checking "the rest of the world".
// They are not the ones the NN was trained on.
/* {  1,  0,  0,  0,  0,  1,        //  'X'
   0,  1,  0,  0,  1,  0,
   0,  0,  1,  1,  0,  0,
   0,  0,  1,  1,  0,  0,
   0,  1,  0,  0,  1,  0,
   1,  0,  0,  0,  0,  1},
{  0,  1,  0,  0,  0,  1,          //  'Y'
   0,  0,  1,  0,  1,  0,
   0,  0,  0,  1,  0,  0,
   0,  0,  0,  1,  0,  0,
   0,  0,  0,  1,  0,  0,
   0,  0,  0,  1,  0,  0},
{  1,  1,  1,  1,  1,  1,          //  'Z'
   0,  0,  0,  0,  1,  0,
   0,  0,  0,  1,  0,  0,
   0,  0,  1,  0,  0,  0,
   0,  1,  0,  0,  0,  0,
   1,  1,  1,  1,  1,  1}*/
},
datatrue[N_DATASETS][N_OUTPUTS] = {{0,1}, {1,0}, {1,1},
```

```
{0,0}, {0,0}, {0,0}, {0,0}, {0,0}, {0,0}};
// Memory allocation and initialization function void MemAllocAndInit(char S)
{
if(S == 'A')
   for(int i = 0; i < N_LAYERS; i++)
   {
w[i] = new float*[conf[i + 1]];
z[i] = new float[conf[i + 1]];


y[i] = new float[conf[i + 1]]; Fi[i] = new float[conf[i + 1]];
for(int j = 0; j < conf[i + 1]; j++)
{
}
}
w[i][j] = new float[conf[i] + 1];
// Initializing in the range (-0.5;0.5) (including bias weight)
for(int k = 0; k <= conf[i]; k++)
  w[i][j][k] = rand()/(float)RAND_MAX - 0.5;
if(S == 'D')
{
for(int i = 0; i < N_LAYERS; i++)
{
}
for(int j = 0; j < conf[i + 1]; j++)
   delete[] w[i][j];
delete[] w[i], z[i], y[i], Fi[i];
}
}
ErrorFile.close();
// Activation function float FNL(float z)
{
}
float y;
y = 1. / (1. + exp(-z));
return y;
// Applying input
void ApplyInput(short sn)
{
float input;
for(short i = 0; i < N_LAYERS; i++) // Counting layers
   for(short j = 0; j < conf[i + 1]; j++) // Counting neurons in each layer
   {
z[i][j] = 0.;

// Counting input to each layer (= # of neurons in the previous layer)
for(short k = 0; k < conf[i]; k++)
{
}
if(i) // If the layer is not the first one input = y[i - 1][k];
else
   input = dataset[sn][k];
z[i][j] += w[i][j][k] * input;
}
}
```

```
z[i][j] += w[i][j][conf[i]]; // Bias term y[i][j] = FNL(z[i][j]);
// Training function, tr - # of runs void Train(int tr)
{
short i, j, k, m, sn;
float eta, prev_output, multiple3, SqErr, eta0;
eta0 = 1.5; // Starting learning rate eta = eta0;
for(m = 0; m < tr; m++) // Going through all tr training runs
{
SqErr = 0.;
// Each training run consists of runs through each training set for(sn = 0;
sn < N_DATASETS; sn++)
{
ApplyInput(sn);
// Counting the layers down
for(i = N_LAYERS - 1; i >= 0; i--)
// Counting neurons in the layer for(j = 0; j < conf[i + 1]; j++)
{
if(i == 2) // If it is the output layer multiple3 = datatrue[sn][j] - y[i][j];
else
{

}
multiple3 = 0.;
// Counting neurons in the following layer for(k = 0; k < conf[i + 2]; k++)
   multiple3 += Fi[i + 1][k] * w[i + 1][k][j];
Fi[i][j] = y[i][j] * (1 - y[i][j]) * multiple3;
// Counting weights in the neuron
// (neurons in the previous layer)
for(k = 0; k < conf[i]; k++)
{
}
if(i) // If it is not a first layer prev_output = y[i - 1][k];
else
   prev_output = dataset[sn][k];
w[i][j][k] += eta * Fi[i][j] * prev_output;
}
// Bias weight correction w[i][j][conf[i]] += eta * Fi[i][j];
}
SqErr += pow((y[N_LAYERS - 1][0] - datatrue[sn][0]), 2) +
   pow((y[N_LAYERS - 1][1] - datatrue[sn][1]), 2);
}
}
ErrorFile << 0.5 * SqErr << endl;
// Decrease learning rate every 100th iteration if(!(m % 100))
   eta /= 2.;
// Go back to original learning rate every 400th iteration if(!(m % 400))
eta = eta0;
// Prints complete information about the network void PrintInfo(void)
{
for(short i = 0; i < N_LAYERS; i++) // Counting layers
{
cout << "LAYER " << i << endl;

// Counting neurons in each layer for(short j = 0; j < conf[i + 1]; j++)
{
```

```
}
}
}
cout << "NEURON " << j << endl;
// Counting input to each layer (= # of neurons in the previous layer)
for(short k = 0; k < conf[i]; k++)
cout << "w[" << i << "][" << j << "][" << k << "]=" << w[i][j][k]
     << ' ';
cout << "w[" << i << "][" << j << "][BIAS]=" << w[i][j][conf[i]]
     << ' ' << endl;
cout << "z[" << i << "][" << j << "]=" << z[i][j] << endl;
cout << "y[" << i << "][" << j << "]=" << y[i][j] << endl;
// Prints the output of the network void PrintOutput(void)
{
// Counting number of datasets
for(short sn = 0; sn < N_DATASETS; sn++)
{
}
}
ApplyInput(sn);
cout << "TRAINING SET " << sn << ": [ ";
// Counting neurons in the output layer for(short j = 0; j < conf[3]; j++)
    cout << y[N_LAYERS - 1][j] << ' ';
cout << "] ";
if(y[N_LAYERS - 1][0] > (datatrue[sn][0] - 0.1)
     && y[N_LAYERS - 1][0] < (datatrue[sn][0] + 0.1)
     && y[N_LAYERS - 1][1] > (datatrue[sn][1] - 0.1)
     && y[N_LAYERS - 1][1] < (datatrue[sn][1] + 0.1))
     cout << "--- RECOGNIZED ---";
else
     cout << "--- NOT RECOGNIZED ---";
cout << endl;

// Loads weithts from a file void LoadWeights(void)
{
float in;
ifstream file("weights.txt", ios::in);
// Counting layers
for(short i = 0; i < N_LAYERS; i++)
      // Counting neurons in each layer
      for(short j = 0; j < conf[i + 1]; j++)
        // Counting input to each layer (= # of neurons in the previous layer)
        for(short k = 0; k <= conf[i]; k++)
        {
}
file >> in;
w[i][j][k] = in;
}
file.close();
// Saves weithts to a file void SaveWeights(void)
{
}
ofstream file("weights.txt", ios::out);
// Counting layers
for(short i = 0; i < N_LAYERS; i++)
```

```
   // Counting neurons in each layer
for(short j = 0; j < conf[i + 1]; j++)
      // Counting input to each layer (= # of neurons in the previous layer)
      for(short k = 0; k <= conf[i]; k++)
         file << w[i][j][k] << endl;
file.close();
// Gathers recognition statistics for 1 and 2 false bit cases void
GatherStatistics(void)
{
short sn, j, k, TotalCases;
int cou;


cout << "WITH 1 FALSE BIT PER CHARACTER:" << endl; TotalCases = conf[0];
// Looking at each dataset
for(sn = 0; sn < N_DATASETS; sn++)
{
cou = 0;
// Looking at each bit in a dataset for(j = 0; j < conf[0]; j++)
{
}
if(dataset[sn][j])
   dataset[sn][j] = 0;
else
   dataset[sn][j] = 1; ApplyInput(sn);
if(y[N_LAYERS - 1][0] > (datatrue[sn][0] - 0.1)
   && y[N_LAYERS - 1][0] < (datatrue[sn][0] + 0.1)
   && y[N_LAYERS - 1][1] > (datatrue[sn][1] - 0.1)
   && y[N_LAYERS - 1][1] < (datatrue[sn][1] + 0.1))
   cou++;
if(dataset[sn][j]) // Switching back dataset[sn][j] = 0;
else
   dataset[sn][j] = 1;
}
cout << "TRAINING SET " << sn << ": " << cou << '/' << TotalCases
     << " recognitions (" << (float)cou / TotalCases * 100. << "%)"
     << endl;
cout << "WITH 2 FALSE BITS PER CHARACTER:" << endl; TotalCases =
conf[0] * (conf[0] - 1.);
// Looking at each dataset
for(sn = 0; sn < N_DATASETS; sn++)
{
cou = 0;
// Looking at each bit in a dataset for(j = 0; j < conf[0]; j++)


for(k = 0; k < conf[0]; k++)
{
}
if(j == k)
   continue;
if(dataset[sn][j])
   dataset[sn][j] = 0;
else
   dataset[sn][j] = 1;
```

```
if(dataset[sn][k])
   dataset[sn][k] = 0;
else
   dataset[sn][k] = 1; ApplyInput(sn);
if(y[N_LAYERS - 1][0] > (datatrue[sn][0] - 0.1)
   && y[N_LAYERS - 1][0] < (datatrue[sn][0] + 0.1)
   && y[N_LAYERS - 1][1] > (datatrue[sn][1] - 0.1)
   && y[N_LAYERS - 1][1] < (datatrue[sn][1] + 0.1))
   cou++;
if(dataset[sn][j]) // Switching back dataset[sn][j] = 0;
else
   dataset[sn][j] = 1;
if(dataset[sn][k])
   dataset[sn][k] = 0;
else
   dataset[sn][k] = 1;
}
}
cout << "TRAINING SET " << sn << ": " << cou << '/' << TotalCases
     << " recognitions (" << (float)cou / TotalCases * 100. << "%)"
     << endl;
// Entry point: main menu void main(void)
{
short ch;
int x;


MemAllocAndInit('A');
do
{
system("cls");
cout << "MENU" << endl;
cout << "1. Apply input and print parameters" << endl;
cout << "2. Apply input (all training sets) and print output" << endl;
cout << "3. Train network" << endl; cout << "4. Load weights" << endl;
cout << "5. Save weights" << endl;
cout << "6. Gather recognition statistics" << endl;
cout << "0. Exit" << endl; cout << "Your choice: "; cin >> ch;
cout << endl;
switch(ch)
{
case 1: cout << "Enter set number: ";
  cin >> x;
  ApplyInput(x);
  PrintInfo(); break;
case 2: PrintOutput();
  break;
case 3: cout << "How many training runs?: ";
  cin >> x; Train(x); break;
case 4: LoadWeights();
  break;
case 5: SaveWeights();
  break;
case 6: GatherStatistics();
  break;
```

```
case 0: MemAllocAndInit('D');
  return;


}
}
cout << endl;
cin.get();
cout << "Press ENTER to continue..." << endl;
cin.get();
}
while(ch);
```

6.B. Back Propagation Case Study[†]: The Exclusive-OR (XOR) Problem (2-Layer BP)

The final weights and outputs for XOR *2 layers* Network after 200 iterations are

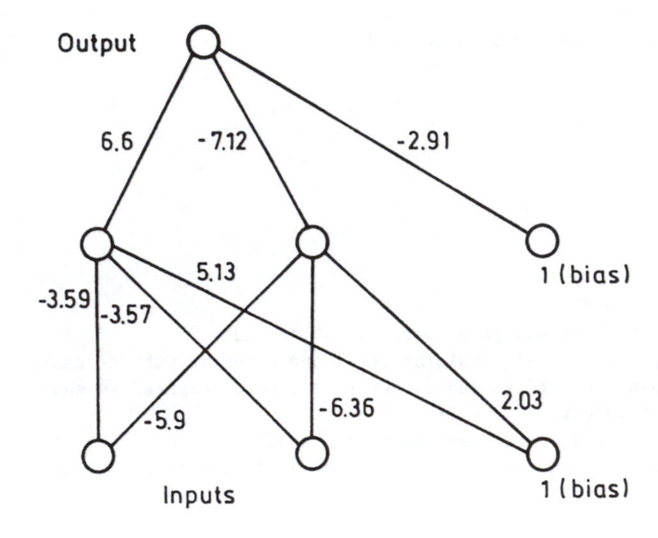

6.B.1. Final weight values.

$$\begin{aligned}\text{input:}\quad &(0,0) \to (0.06) = (\text{output})\\ &(0,1) \to (0.91)\\ &(1,0) \to (0.91)\\ &(1,1) \to (0.11)\end{aligned}$$

Starting learning rate: 6
Learning rate after 100 iterations: 3

[†]Computed by Mr. Sang Lee, EECS Dept., University of Illinois, Chicago, 1993.

The C-language program for the aboves XOR problem is as follows:

```
/***************************************************************************
 ***

    PROGRAM: XOR2.c

    PURPOSE: Approximating Exclusive-Or function using Neural Network with
             2 Hidden Layers.

    FUNCTIONS:

        WinMain() - calls initialization function, processes message loop
        InitApplication() - initializes window data and registers window
        InitInstance() - saves instance handle and creates main window
        MainWndProc() - processes messages
        About() - processes messages for "About" dialog box

****************************************************************************
 **/

#include <windows.h>
#include <dde.h>
#include <io.h>
#include <time.h>
#include <string.h>
#include <stdlib.h>
#include <math.h>
#include "xor2.h"

HANDLE hInst;

/***************************************************************************
 ***

    FUNCTION: WinMain(HANDLE, HANDLE, LPSTR, int)

    PURPOSE: calls initialization function, processes message loop

****************************************************************************
 **/

int PASCAL WinMain(hInstance, hPrevInstance, lpCmdLine, nCmdShow)
HANDLE hInstance;
HANDLE hPrevInstance;
LPSTR lpCmdLine;
int nCmdShow;
{
    MSG msg;

    if (!hPrevInstance)
        if (!InitApplication(hInstance))
            return (FALSE);

    if (!InitInstance(hInstance, nCmdShow))
        return (FALSE);
```

```
                          XOR2.C
                                    InitApplication

 !    ——while (GetMessage(&msg. NULL, NULL, NULL)) {
 |    |      TranslateMessage(&msg);
 |    |      DispatchMessage(&msg);
 |    └——}
 |
 |
 |       return (msg.wParam);
 └——}

   /**************************************************************************
    ***

       FUNCTION: InitApplication(HANDLE)

       PURPOSE: Initializes window data and registers window class

   **************************************************************************
    **/

   BOOL InitApplication(hInstance)
   HANDLE hInstance;
 ┌——{
 |      WNDCLASS  wc;
 |
 |      wc.style = NULL;
 |      wc.lpfnWndProc = MainWndProc;
 |      wc.cbClsExtra = 0;
 |      wc.cbWndExtra = 0;
 |      wc.hInstance = hInstance;
 |      wc.hIcon = LoadIcon(hInstance, "xor");            /* loads icon */
 |      wc.hCursor = LoadCursor(NULL. IDC_ARROW);
 |      wc.hbrBackground = GetStockObject(WHITE_BRUSH);
 |      wc.lpszMenuName =  "BallMenu";
 |      wc.lpszClassName = "XorWClass";
 |
 |      return (RegisterClass(&wc));
 └——}

   /**************************************************************************
    ***

       FUNCTION:  InitInstance(HANDLE, int)

       PURPOSE:  Saves instance handle and creates main window

   **************************************************************************
    **/

   BOOL InitInstance(hInstance, nCmdShow)
   HANDLE            hInstance;
   int               nCmdShow;
 ┌——{
 |      HWND              hWnd;
```

XOR2.C
MainWndProc

```
    LONG            result;

    hInst = hInstance;

    hWnd = CreateWindow(
    "XorWClass",
    "XOR with 2 Layers",
    WS_OVERLAPPEDWINDOW,
    0,
    240,
    640,
    240,
    NULL,
    NULL,
    hInstance,
    NULL
    );


    if (!hWnd)
        return (FALSE);

    ShowWindow(hWnd, nCmdShow);
    UpdateWindow(hWnd);
    return (TRUE);

}

/****************************************************************************
 ***

    FUNCTION: MainWndProc(HWND, unsigned, WORD, LONG)

    PURPOSE:  Processes messages

    MESSAGES:

        WM_COMMAND    - application menu (About dialog box)
        WM_DESTROY    - destroy window

****************************************************************************
 **/

long FAR PASCAL MainWndProc(hWnd, message, wParam, lParam)
HWND hWnd;
unsigned message;
WORD wParam;
LONG lParam;
{
    HDC hdc,hdcPrn;
    RECT rect;
    PAINTSTRUCT ps;
    OFSTRUCT OfStruct;
    int fAckReq;
    int hFile;
```

XOR2.C
MainWndProc

```
static int count;
int i,j,k,index,indexi,pcount;
static WEIGHTSTR weight;
char out[300];
/* to support clip board message passing */
HANDLE hText,hMyText;
    /* to store result */
static float first,second,third,fourth,learnrate;
float delta[NUNITS];
float error[NUNITS];
float net_input[NUNITS];
float activ[NUNITS];
float target[NOUTPUTS];
    /* This is the colors for outputs */
static LOGPEN lpBlack = {PS_SOLID,1,1,RGB(0,0,0)},
lpBlue = {PS_SOLID,1,1,RGB(0,0,255)},
lpYellow = {PS_SOLID,1,1,RGB(150,150,0)},
lpRed = {PS_SOLID,1,1,RGB(255,0,0)},
lpGreen = {PS_SOLID,1,1,RGB(0,255,0)};
static HPEN hPenBlack,hPenBlue,hPenYellow,hPenRed,hPenGreen;

switch (message) {

    case WM_CREATE:

                /* initialize weights */
        for (k = 0; k< NHUNITS*NOUTPUTS; k++)
        {
            weight.weight1[k] = rand()%10;
            if (weight.weight1[k] > 5)
                weight.weight1[k] = -weight.weight1[k];
            weight.weight1[k] = weight.weight1[k]/10;
        }
        for (k = 0; k< NHUNITS*NINPUTS; k++)
        {
            weight.weight0[k] = rand()%10;
            if (weight.weight0[k] > 5)
                weight.weight0[k] = -weight.weight0[k];
            weight.weight0[k] = weight.weight0[k]/10;
        }

                /* initialize learing rate . this is going to be
                   smaller as time goes on */
        learnrate = 5;
        count = 0;
                /* initialize timer */
        if(!SetTimer(hWnd,ID_TIMER,500,NULL))
        {
            MessageBox(hWnd,"Too many clocks or timers !","Ball",
            MB_ICONEXCLAMATION | MB_OK);
            return FALSE;
        }

                /* initialize color pens */
```

```
XOR2.C
                MainWndProc

        hPenBlack = CreatePenIndirect(&lpBlack);
        hPenBlue = CreatePenIndirect(&lpBlue);
        hPenYellow = CreatePenIndirect(&lpYellow);
        hPenRed = CreatePenIndirect(&lpRed);
        hPenGreen = CreatePenIndirect(&lpGreen);
        return 0;

    case WM_TIMER:

             /* decrease learning rate to have fast convergence */
        if ((count%101) == 100)
            learnrate = learnrate/2;

                    /****************************************/
                    /* put first input into neural network */
                    /****************************************/
        activ[0] = 0;
        activ[1] = 0;
                    /* This is bias */
        activ[NINPUTS-1] = 1;

                    /**********************************/
                    /* forward activation propagation */
                    /**********************************/

                    /* calculate activation for hidden nodes */
        for (i = NINPUTS,k=0; i< NINPUTS+NHUNITS;i++)
        {
            net_input[i] = 0;
            for(j = 0; j< NINPUTS;j++,k++)
            {
                net_input[i] += activ[j]*weight.weight0[k];
            }
                      /* apply activation function */
            activ[i] = 1/(1+(float)exp(-net_input[i])) ;
        }
                    /* This is bias */
        activ[NINPUTS+NHUNITS-1] = 1;

                    /* calculate activation for output nodes */
        for (i = NINPUTS+NHUNITS,k=0; i< NINPUTS+NHUNITS+NOUTPUTS;i++)
        {
            net_input[i] = 0;
            for(j = NINPUTS; j< NINPUTS+NHUNITS;j++,k++)
            {
                net_input[i] += activ[j]*weight.weight1[k];
            }
                      /* apply activation function */
            activ[i] = 1/(1+(float)exp(-net_input[i])) ;
        }

                    /* this is final activation of the neural network
                    */
        first = activ[NUNITS-1];
```

```
                    XOR2.C
                              MainWndProc

              |      error[i] = target[k] - activ[i];
              └─}

                         /***********************************************
                         */
                         /* backward error propagation for the third input
                         */
                         /***********************************************/

                         /* initialize errors before doing something else */
          for (i = 0; i < NUNITS-1; i++)
              ┌─{
              |      error[i] = 0;
              └─}

                         /* calculate errors between outputs and hidden
                          nodes  */
          for ( i = NUNITS-NOUTPUTS,k=0; i < NUNITS; i++)
              ┌─{
              |      delta[i] = error[i]*activ[i]*(1.0 -activ[i]);
              |                                              /* error
              |                                  back propagate before
              |                                  bias */
              |      for (j = NUNITS-NOUTPUTS-NHUNITS; j < NUNITS-NOUTPUTS-1;
              |        j++,k++)
              |          error[j] += delta[i]*weight.weight1[k];
              └─}

                         /* calculate errors between hidden nodes  and
                          input nodes*/
          for ( i = NUNITS-NOUTPUTS-NHUNITS,k=0; i < NUNITS-NOUTPUTS-1;
            i++)
              ┌─{
              |      delta[i] = error[i]*activ[i]*(1.0 -activ[i]);
              |              /* we don't need to calculate errors for input
              |               nodes */
              └─}

                         /* calculate delta weight changes between outputs
                          and hidden nodes  */
          for ( i = NUNITS-NOUTPUTS,k=0; i < NUNITS; i++)
              ┌─{
              |      for (j = NUNITS-NOUTPUTS-NHUNITS; j < NUNITS-NOUTPUTS;
              |        j++,k++)
              |          weight.weight1[k] += delta[i]*activ[j]*learnrate;
              └─}

                         /* calculate delta weight changes between hidden
                          nodes and input nodes*/
          for ( i = NUNITS-NOUTPUTS-NHUNITS,k=0; i < NUNITS-NOUTPUTS;
            i++)
              ┌─{
              |      for (j = 0; j < NINPUTS; j++,k++)
              |          weight.weight0[k] += delta[i]*activ[j]*learnrate;
              └─}
```

```
XOR2.C
                MainWndProc

                /****************************************/
                /* put fourth input into neural network */
                /****************************************/

    activ[0] = 1;
    activ[1] = 1;
                /* This is bias */
    activ[NINPUTS-1] = 1;

                /************************************/
                /* forward activation propagation */
                /************************************/

                /* calculate activation for hidden nodes */
    for (i = NINPUTS,k=0; i< NINPUTS+NHUNITS;i++)
    {
        net_input[i] = 0;
        for(j = 0; j< NINPUTS;j++,k++)
        {
            net_input[i] += activ[j]*weight.weight0[k];
        }
                /* apply activation function */
        activ[i] = 1/(1+(float)exp(-net_input[i])) ;
    }
                /* This is bias */
    activ[NINPUTS+NHUNITS-1] = 1;

                /* calculate activation for output nodes */
    for (i = NINPUTS+NHUNITS,k=0; i< NINPUTS+NHUNITS+NOUTPUTS;i++)
    {
        net_input[i] = 0;
        for(j = NINPUTS; j< NINPUTS+NHUNITS;j++,k++)
        {
            net_input[i] += activ[j]*weight.weight1[k];
        }
                /* apply activation function */
        activ[i] = 1/(1+(float)exp(-net_input[i])) ;
    }
                /* this is final activation of the neural network
                 */
    fourth = activ[NUNITS-1];

                /* using target output, calculate weight changes */
                /* If both inputs are one, then output should be
                 zero */
    target[0] = 0;

/* calculate errors for outputs */
    for (i = NUNITS-NOUTPUTS,k=0; i < NUNITS; i++,k++)
    {
        error[i] = target[k] - activ[i];
    }

                /*************************************************
                 */
```

XOR2.C
MainWndProc

```
|    |          target[0] = 1;
|    |
|    |       /* calculate errors for outputs */
|    |          for (i = NUNITS-NOUTPUTS,k=0; i < NUNITS; i++,k++)
|    |       ┌──{
|    |       |      error[i] = target[k] - activ[i];
|    |       └──}
|    |
|    |                    /**************************************************
|    |                     */
|    |                    /* backward error propagation for the second
|    |                     input */
|    |                    /**************************************************/
|    |
|    |                    /* initialize errors before doing something else */
|    |          for (i = 0; i < NUNITS-1; i++)
|    |       ┌──{
|    |       |      error[i] = 0;
|    |       └──}
|    |
|    |                    /* calculate errors between outputs and hidden
|    |                     nodes  */
|    |          for ( i = NUNITS-NOUTPUTS,k=0; i < NUNITS; i++)
|    |       ┌──{
|    |       |      delta[i] = error[i]*activ[i]*(1.0 -activ[i]);
|    |       |                                                   /* error
|    |       |                                        back propagate before
|    |       |                                        bias */
|    |       |      for (j = NUNITS-NOUTPUTS-NHUNITS; j < NUNITS-NOUTPUTS-1;
|    |       |        j++,k++)
|    |       |          error[j] += delta[i]*weight.weight1[k];
|    |       └──}
|    |
|    |                    /* calculate errors between hidden nodes  and
|    |                     input nodes*/
|    |          for ( i = NUNITS-NOUTPUTS-NHUNITS,k=0; i < NUNITS-NOUTPUTS-1;
|    |            i++)
|    |       ┌──{
|    |       |      delta[i] = error[i]*activ[i]*(1.0 -activ[i]);
|    |       |                /* we don't need to calculate errors for input
|    |       |                 nodes */
|    |       └──}
|    |
|    |                    /* calculate delta weight changes between outputs
|    |                     and hidden nodes  */
|    |          for ( i = NUNITS-NOUTPUTS,k=0; i < NUNITS; i++)
|    |       ┌──{
|    |       |      for (j = NUNITS-NOUTPUTS-NHUNITS; j < NUNITS-NOUTPUTS;
|    |       |        j++,k++)
|    |       |          weight.weight1[k] += delta[i]*activ[j]*learnrate;
|    |       └──}
|    |
|    |                    /* calculate delta weight changes between hidden
|    |                     nodes and input nodes*/
|    |          for ( i = NUNITS-NOUTPUTS-NHUNITS,k=0; i < NUNITS-NOUTPUTS;
```

XOR2.C
MainWndProc

```
          i++)
       {
          for (j = 0; j < NINPUTS; j++,k++)
              weight.weight0[k] += delta[i]*activ[j]*learnrate;
       }
                    /*************************************/
                    /* put 3rd input into neural network */
                    /*************************************/

   activ[0] = 1;
   activ[1] = 0;
                    /* This is bias */
   activ[NINPUTS-1] = 1;

                    /**********************************/
                    /* forward activation propagation */
                    /**********************************/

                    /* calculate activation for hidden nodes */
   for (i = NINPUTS,k=0; i< NINPUTS+NHUNITS;i++)
   {
       net_input[i] = 0;
       for(j = 0; j< NINPUTS;j++,k++)
       {
           net_input[i] += activ[j]*weight.weight0[k];
       }
                    /* apply activation function */
       activ[i] = 1/(1+(float)exp(-net_input[i])) ;
   }
                    /* This is bias */
   activ[NINPUTS+NHUNITS-1] = 1;

                    /* calculate activation for output nodes */
   for (i = NINPUTS+NHUNITS,k=0; i< NINPUTS+NHUNITS+NOUTPUTS;i++)
   {
       net_input[i] = 0;
       for(j = NINPUTS; j< NINPUTS+NHUNITS;j++,k++)
       {
           net_input[i] += activ[j]*weight.weight1[k];
       }
                    /* apply activation function */
       activ[i] = 1/(1+(float)exp(-net_input[i])) ;
   }
                    /* this is final activation of the neural network
                    */
   third = activ[NUNITS-1];

                    /* using target output, calculate weight changes */
                    /* If one input is not zero, then output should
                     be one */
   target[0] = 1;

/* calculate errors for outputs */
   for (i = NUNITS-NOUTPUTS,k=0; i < NUNITS; i++,k++)
   {
```

```
                    XOR2.C
                              MainWndProc

                         /* using target output, calculate weight changes */
                         /* If both inputs are zero, then output should be
                          zero. */
        target[0] = 0;

     /* calculate errors for outputs */
        for (i = NUNITS-NOUTPUTS,k=0; i < NUNITS; i++,k++)
     ┌──{
     │      error[i] = target[k] - activ[i];
     └──}

                         /************************************************
                          */
                         /* backward error propagation for the first input
                          */
                         /************************************************
                          */

                         /* initialize errors before doing something else */
        for (i = 0; i < NUNITS-1; i++)
     ┌──{
     │      error[i] = 0;
     └──}

                         /* calculate errors between outputs and hidden
                          nodes  */
        for ( i = NUNITS-NOUTPUTS,k=0; i < NUNITS; i++)
     ┌──{
     │      delta[i] = error[i]*activ[i]*(1.0 -activ[i]);
     │                                                       /* error
     │                                       back propagate before
     │                                       bias */
     │      for (j = NUNITS-NOUTPUTS-NHUNITS; j < NUNITS-NOUTPUTS-1;
     │        j++,k++)
     │          error[j] += delta[i]*weight.weight1[k];
     └──}

                         /* calculate errors between hidden nodes  and
                          input nodes*/
        for ( i = NUNITS-NOUTPUTS-NHUNITS,k=0; i < NUNITS-NOUTPUTS-1;
          i++)
     ┌──{
     │      delta[i] = error[i]*activ[i]*(1.0 -activ[i]);
     │                   /* we don't need to calculate errors for input
     │                    nodes */
     └──}

                         /* calculate delta weight changes between outputs
                          and hidden nodes  */
        for ( i = NUNITS-NOUTPUTS,k=0; i < NUNITS; i++)
     ┌──{
     │      for (j = NUNITS-NOUTPUTS-NHUNITS; j < NUNITS-NOUTPUTS;
     │        j++,k++)
     │          weight.weight1[k] += delta[i]*activ[j]*learnrate;
     └──}
```

XOR2.C
MainWndProc

```
            /* calculate delta weight changes between hidden
             nodes and input nodes*/
for ( i = NUNITS-NOUTPUTS-NHUNITS,k=0; i < NUNITS-NOUTPUTS;
  i++)
{
    for (j = 0; j < NINPUTS; j++,k++)
        weight.weight0[k] += delta[i]*activ[j]*learnrate;
}

            /******************************************/
            /* put second input into neural network   */
            /******************************************/
activ[0] = 0;
activ[1] = 1;
            /* This is bias */
activ[NINPUTS-1] = 1;

            /**********************************/
            /* forward activation propagation */
            /**********************************/

            /* calculate activation for hidden nodes */
for (i = NINPUTS,k=0; i< NINPUTS+NHUNITS;i++)
{
    net_input[i] = 0;
    for(j = 0; j< NINPUTS;j++,k++)
    {
        net_input[i] += activ[j]*weight.weight0[k];
    }
            /* apply activation function */
    activ[i] = 1/(1+(float)exp(-net_input[i])) ;
}
            /* This is bias */
activ[NINPUTS+NHUNITS-1] = 1;

            /* calculate activation for output nodes */
for (i = NINPUTS+NHUNITS,k=0; i< NINPUTS+NHUNITS+NOUTPUTS;i++)
{
    net_input[i] = 0;
    for(j = NINPUTS; j< NINPUTS+NHUNITS;j++,k++)
    {
        net_input[i] += activ[j]*weight.weight1[k];
    }
            /* apply activation function */
    activ[i] = 1/(1+(float)exp(-net_input[i])) ;
}

            /* this is final activation of the neural network
             */
second = activ[NUNITS-1];

            /* using target output, calculate weight changes */
            /* If one of the inputs is not zero, then output
             should be one */
```

```
XOR2.C
                    MainWndProc

                /* backward error propagation for the fourth
                 input */
                /***********************************************/

                /* initialize errors before doing something else */
    for (i = 0; i < NUNITS-1; i++)
    {
        error[i] = 0;
    }

                /* calculate errors between outputs and hidden
                 nodes  */
    for ( i = NUNITS-NOUTPUTS,k=0; i < NUNITS; i++)
    {
        delta[i] = error[i]*activ[i]*(1.0 -activ[i]);
                    /* error back propagate before bias */
        for (j = NUNITS-NOUTPUTS-NHUNITS; j < NUNITS-NOUTPUTS-1;
          j++,k++)
            error[j] += delta[i]*weight.weight1[k];
    }

                /* calculate errors between hidden nodes  and
                 input nodes*/
    for ( i = NUNITS-NOUTPUTS-NHUNITS,k=0; i < NUNITS-NOUTPUTS-1;
      i++)
    {
        delta[i] = error[i]*activ[i]*(1.0 -activ[i]);
                /* we don't need to calculate errors for input
                 nodes */
    }

                /* calculate delta weight changes between outputs
                 and hidden nodes  */
    for ( i = NUNITS-NOUTPUTS,k=0; i < NUNITS; i++)
    {
        for (j = NUNITS-NOUTPUTS-NHUNITS; j < NUNITS-NOUTPUTS;
          j++,k++)
            weight.weight1[k] += delta[i]*activ[j]*learnrate;
    }

                /* calculate delta weight changes between hidden
                 nodes and input nodes*/
    for ( i = NUNITS-NOUTPUTS-NHUNITS,k=0; i < NUNITS-NOUTPUTS;
      i++)
    {
        for (j = 0; j < NINPUTS; j++,k++)
            weight.weight0[k] += delta[i]*activ[j]*learnrate;
    }

                /* Now save the result in the file */
    hFile = OpenFile("xor2.fil",&OfStruct,OF_EXIST);
    if (hFile >= 0)
    {
        hFile = OpenFile("xor2.fil",&OfStruct,OF_READWRITE);
        if (hFile < 0)
```

```
                XOR2.C
                         MainWndProc

|     |          |    ┌──{
|     |          |    └──}
|     |          └──} /* end of exist file */
|     |             else
|     |          ┌──{
|     |          |       hFile = OpenFile("xor2.fil",&OfStruct,OF_CREATE);
|     |          |       if (hFile < 0)
|     |          |    ┌──{
|     |          |    └──}
|     |          └──} /* end of created error file */
|     |             i = 0;
|     |             write(hFile,(char*)&count,2);
|     |             while(i < count)
|     |          ┌──{
|     |          |                /* 16 + WEIGHT STRUCT SIZE = 40 */
|     |          |       read(hFile,out,16+sizeof(WEIGHTSTR));
|     |          |       i++;
|     |          └──}
|     |             write(hFile,(char*)&first,4);
|     |             write(hFile,(char*)&second,4);
|     |             write(hFile,(char*)&third,4);
|     |             write(hFile,(char*)&fourth,4);
|     |                         /* write weights */
|     |             write(hFile,(char*)&weight,sizeof(WEIGHTSTR));
|     |                         /* Increase count and iteration # */
|     |             count++;
|     |             close(hFile);
|     |             InvalidateRect(hWnd,NULL,TRUE);
|     |             break;
|     |
|     |        case WM_PAINT:
|     |             GetClientRect(hWnd,&rect);
|     |             hdc = BeginPaint(hWnd,&ps);
|     |             SelectObject(hdc,hPenBlack);
|     |
|     |             wsprintf(out,"Count# %d LearnR %ld.%02ld Blue(0,0:0.%02ld)
|     |              Yellow(0,1:0.%02ld) Red(1,0:0.%02ld) Green(1,1:0.%02ld)",
|     |             count,(long)learnrate,((long)(learnrate*100))%100,
|     |             (((long)(first*100))%100),(((long)(second*100))%100),(((long)(
|     |              third*100))%100),(((long)(fourth*100))%100)
|     |             );
|     |             DrawText(hdc,out,-1,&rect,DT_CENTER);
|     |             Rectangle(hdc,10,10,15,210);
|     |
|     |             SelectObject(hdc,hPenBlue);
|     |             Rectangle(hdc,200, 10+(200 - (int)(((long)(first*100))%100)*2)
|     |              ,210,210);
|     |
|     |             SelectObject(hdc,hPenYellow);
|     |             Rectangle(hdc,300, 10+(200 - (int)(((long)(second*100))%100)*
|     |              2),310,210);
|     |
|     |             SelectObject(hdc,hPenRed);
|     |             Rectangle(hdc,400, 10+(200 - (int)(((long)(third*100))%100)*2)
|     |              ,410,210);
```

```
                    XOR2.C
                              MainWndProc

|    |
|    |          SelectObject(hdc,hPenGreen);
|    |          Rectangle(hdc,500, 10+(200 - (int)(((long)(fourth*100))%100)*
|    |           2),510,210);
|    |
|    |          EndPaint(hWnd,&ps);
|    |          break;
|    |      case WM_DESTROY:
|    |          PostQuitMessage(0);
|    |          break;
|    |
|    |      default:
|    |          return (DefWindowProc(hWnd, message, wParam, lParam));
|    └──}
|       return (NULL);
└──}
```

Computation Results (see footnote at end of table for notation)*

```
Training Neural Network for XOR function with 2 Layers
 [# 1] (0 0 0.51) (0 1 0.27) (1 0 0.55) (1 1 0.76)
  W Out to H (0.07,-0.67,0.34) H to I 1(-0.07,-1.04,0.24) 2(-0.87,-0.98,0.04)
 [# 2] (0 0 0.50) (0 1 0.29) (1 0 0.54) (1 1 0.75)
  W Out to H (0.05,-0.67,0.30) H to I 1(-0.13,-1.15,0.10) 2(-0.94,-1.14,-0.10)
 [# 3] (0 0 0.50) (0 1 0.31) (1 0 0.54) (1 1 0.75)
  W Out to H (0.03,-0.69,0.27) H to I 1(-0.18,-1.23,-0.02) 2(-1.00,-1.27,-0.22
 [# 4] (0 0 0.49) (0 1 0.32) (1 0 0.54) (1 1 0.74)
  W Out to H (0.01,-0.74,0.26) H to I 1(-0.22,-1.30,-0.12) 2(-1.08,-1.39,-0.31
 [# 5] (0 0 0.48) (0 1 0.32) (1 0 0.54) (1 1 0.74)
  W Out to H (-0.00,-0.79,0.25) H to I 1(-0.26,-1.36,-0.20) 2(-1.15,-1.50,-0.3
 [# 6] (0 0 0.48) (0 1 0.33) (1 0 0.54) (1 1 0.74)
  W Out to H (-0.02,-0.86,0.25) H to I 1(-0.29,-1.41,-0.28) 2(-1.22,-1.60,-0.4
 [# 7] (0 0 0.47) (0 1 0.33) (1 0 0.54) (1 1 0.74)
  W Out to H (-0.04,-0.94,0.25) H to I 1(-0.32,-1.45,-0.35) 2(-1.29,-1.69,-0.4
 [# 8] (0 0 0.46) (0 1 0.34) (1 0 0.55) (1 1 0.74)
  W Out to H (-0.06,-1.02,0.26) H to I 1(-0.36,-1.49,-0.41) 2(-1.37,-1.78,-0.5
 [# 9] (0 0 0.46) (0 1 0.35) (1 0 0.55) (1 1 0.74)
  W Out to H (-0.09,-1.11,0.27) H to I 1(-0.39,-1.53,-0.46) 2(-1.44,-1.87,-0.5
 [# 10] (0 0 0.45) (0 1 0.35) (1 0 0.55) (1 1 0.75)
  W Out to H (-0.11,-1.20,0.28) H to I 1(-0.42,-1.56,-0.51) 2(-1.52,-1.96,-0.5
 [# 11] (0 0 0.44) (0 1 0.36) (1 0 0.56) (1 1 0.75)
  W Out to H (-0.14,-1.30,0.30) H to I 1(-0.45,-1.58,-0.55) 2(-1.60,-2.05,-0.5
 [# 12] (0 0 0.44) (0 1 0.37) (1 0 0.57) (1 1 0.75)
  W Out to H (-0.17,-1.41,0.32) H to I 1(-0.48,-1.61,-0.60) 2(-1.68,-2.14,-0.5
 [# 13] (0 0 0.43) (0 1 0.38) (1 0 0.57) (1 1 0.76)
  W Out to H (-0.20,-1.52,0.35) H to I 1(-0.51,-1.63,-0.63) 2(-1.76,-2.23,-0.5
 [# 14] (0 0 0.42) (0 1 0.38) (1 0 0.58) (1 1 0.76)
  W Out to H (-0.23,-1.63,0.38) H to I 1(-0.53,-1.66,-0.66) 2(-1.85,-2.33,-0.4
 [# 15] (0 0 0.41) (0 1 0.39) (1 0 0.59) (1 1 0.76)
  W Out to H (-0.26,-1.75,0.41) H to I 1(-0.56,-1.68,-0.69) 2(-1.93,-2.43,-0.4
 [# 16] (0 0 0.40) (0 1 0.41) (1 0 0.59) (1 1 0.77)
  W Out to H (-0.29,-1.88,0.44) H to I 1(-0.59,-1.70,-0.72) 2(-2.02,-2.52,-0.3
 [# 17] (0 0 0.39) (0 1 0.42) (1 0 0.60) (1 1 0.77)
  W Out to H (-0.32,-2.00,0.47) H to I 1(-0.62,-1.73,-0.74) 2(-2.11,-2.63,-0.3
 [# 18] (0 0 0.38) (0 1 0.43) (1 0 0.61) (1 1 0.78)
  W Out to H (-0.35,-2.13,0.51) H to I 1(-0.65,-1.75,-0.76) 2(-2.20,-2.73,-0.2
 [# 19] (0 0 0.36) (0 1 0.45) (1 0 0.62) (1 1 0.78)
  W Out to H (-0.37,-2.26,0.55) H to I 1(-0.68,-1.77,-0.78) 2(-2.30,-2.83,-0.1
 [# 20] (0 0 0.35) (0 1 0.46) (1 0 0.63) (1 1 0.79)
  W Out to H (-0.40,-2.39,0.58) H to I 1(-0.71,-1.79,-0.80) 2(-2.39,-2.94,-0.1
 [# 21] (0 0 0.33) (0 1 0.48) (1 0 0.64) (1 1 0.79)
  W Out to H (-0.42,-2.52,0.62) H to I 1(-0.73,-1.80,-0.82) 2(-2.48,-3.04,-0.0
 [# 22] (0 0 0.32) (0 1 0.50) (1 0 0.65) (1 1 0.80)
  W Out to H (-0.43,-2.65,0.65) H to I 1(-0.76,-1.82,-0.83) 2(-2.57,-3.14,0.01
 [# 23] (0 0 0.30) (0 1 0.51) (1 0 0.66) (1 1 0.80)
  W Out to H (-0.45,-2.77,0.68) H to I 1(-0.78,-1.84,-0.84) 2(-2.67,-3.24,0.08
 [# 24] (0 0 0.29) (0 1 0.53) (1 0 0.66) (1 1 0.80)
  W Out to H (-0.46,-2.89,0.71) H to I 1(-0.80,-1.85,-0.85) 2(-2.76,-3.33,0.14
 [# 25] (0 0 0.27) (0 1 0.54) (1 0 0.67) (1 1 0.81)
  W Out to H (-0.46,-3.00,0.73) H to I 1(-0.82,-1.87,-0.87) 2(-2.84,-3.43,0.19
```

```
[# 26] (0 0 0.25) (0 1 0.56) (1 0 0.68) (1 1 0.81)
 W Out to H (-0.47,-3.11,0.75) H to I 1(-0.84,-1.88,-0.88) 2(-2.93,-3.51,0.24
[# 27] (0 0 0.24) (0 1 0.57) (1 0 0.68) (1 1 0.81)
 W Out to H (-0.47,-3.20,0.77) H to I 1(-0.86,-1.89,-0.89) 2(-3.01,-3.60,0.28
[# 28] (0 0 0.23) (0 1 0.58) (1 0 0.69) (1 1 0.81)
 W Out to H (-0.46,-3.29,0.78) H to I 1(-0.88,-1.90,-0.90) 2(-3.09,-3.67,0.31
[# 29] (0 0 0.22) (0 1 0.59) (1 0 0.69) (1 1 0.81)
 W Out to H (-0.46,-3.38,0.78) H to I 1(-0.89,-1.91,-0.91) 2(-3.16,-3.75,0.35
[# 30] (0 0 0.21) (0 1 0.60) (1 0 0.70) (1 1 0.81)
 W Out to H (-0.45,-3.46,0.79) H to I 1(-0.91,-1.92,-0.92) 2(-3.23,-3.82,0.37
[# 31] (0 0 0.20) (0 1 0.60) (1 0 0.70) (1 1 0.81)
 W Out to H (-0.44,-3.53,0.79) H to I 1(-0.92,-1.93,-0.93) 2(-3.30,-3.88,0.39
[# 32] (0 0 0.19) (0 1 0.61) (1 0 0.71) (1 1 0.81)
 W Out to H (-0.42,-3.60,0.79) H to I 1(-0.94,-1.94,-0.95) 2(-3.37,-3.95,0.41
[# 33] (0 0 0.18) (0 1 0.61) (1 0 0.71) (1 1 0.81)
 W Out to H (-0.41,-3.66,0.79) H to I 1(-0.95,-1.94,-0.96) 2(-3.43,-4.01,0.43
[# 34] (0 0 0.17) (0 1 0.62) (1 0 0.71) (1 1 0.81)
 W Out to H (-0.39,-3.72,0.79) H to I 1(-0.96,-1.95,-0.97) 2(-3.49,-4.07,0.44
[# 35] (0 0 0.16) (0 1 0.62) (1 0 0.71) (1 1 0.81)
 W Out to H (-0.38,-3.78,0.78) H to I 1(-0.97,-1.96,-0.98) 2(-3.55,-4.12,0.46
[# 36] (0 0 0.16) (0 1 0.62) (1 0 0.72) (1 1 0.81)
 W Out to H (-0.36,-3.83,0.78) H to I 1(-0.98,-1.96,-0.99) 2(-3.60,-4.17,0.47
[# 37] (0 0 0.15) (0 1 0.62) (1 0 0.72) (1 1 0.81)
 W Out to H (-0.34,-3.88,0.77) H to I 1(-0.99,-1.97,-1.00) 2(-3.66,-4.23,0.47
[# 38] (0 0 0.15) (0 1 0.63) (1 0 0.72) (1 1 0.81)
 W Out to H (-0.32,-3.93,0.77) H to I 1(-1.00,-1.97,-1.01) 2(-3.71,-4.28,0.48
[# 39] (0 0 0.14) (0 1 0.63) (1 0 0.72) (1 1 0.81)
 W Out to H (-0.30,-3.97,0.76) H to I 1(-1.00,-1.98,-1.02) 2(-3.76,-4.32,0.49
[# 40] (0 0 0.14) (0 1 0.63) (1 0 0.72) (1 1 0.81)
 W Out to H (-0.28,-4.01,0.76) H to I 1(-1.01,-1.98,-1.02) 2(-3.81,-4.37,0.49
[# 41] (0 0 0.14) (0 1 0.63) (1 0 0.72) (1 1 0.81)
 W Out to H (-0.26,-4.05,0.75) H to I 1(-1.02,-1.99,-1.03) 2(-3.85,-4.41,0.50
[# 42] (0 0 0.13) (0 1 0.63) (1 0 0.72) (1 1 0.81)
 W Out to H (-0.23,-4.09,0.75) H to I 1(-1.02,-1.99,-1.04) 2(-3.89,-4.46,0.50
[# 43] (0 0 0.13) (0 1 0.63) (1 0 0.72) (1 1 0.81)
 W Out to H (-0.21,-4.13,0.74) H to I 1(-1.03,-1.99,-1.04) 2(-3.94,-4.50,0.51
[# 44] (0 0 0.13) (0 1 0.63) (1 0 0.73) (1 1 0.81)
 W Out to H (-0.19,-4.16,0.73) H to I 1(-1.03,-2.00,-1.05) 2(-3.98,-4.54,0.51
[# 45] (0 0 0.12) (0 1 0.63) (1 0 0.73) (1 1 0.81)
 W Out to H (-0.16,-4.20,0.73) H to I 1(-1.04,-2.00,-1.06) 2(-4.02,-4.58,0.52)
[# 46] (0 0 0.12) (0 1 0.63) (1 0 0.73) (1 1 0.81)
 W Out to H (-0.14,-4.23,0.72) H to I 1(-1.04,-2.00,-1.06) 2(-4.06,-4.61,0.52)
[# 47] (0 0 0.12) (0 1 0.63) (1 0 0.73) (1 1 0.81)
 W Out to H (-0.11,-4.26,0.72) H to I 1(-1.04,-2.00,-1.07) 2(-4.09,-4.65,0.52)
[# 48] (0 0 0.12) (0 1 0.63) (1 0 0.73) (1 1 0.81)
 W Out to H (-0.09,-4.29,0.71) H to I 1(-1.04,-2.01,-1.07) 2(-4.13,-4.69,0.53
[# 49] (0 0 0.11) (0 1 0.63) (1 0 0.73) (1 1 0.81)
 W Out to H (-0.06,-4.32,0.71) H to I 1(-1.05,-2.01,-1.07) 2(-4.17,-4.72,0.53)
[# 50] (0 0 0.11) (0 1 0.63) (1 0 0.73) (1 1 0.80)
 W Out to H (-0.04,-4.35,0.70) H to I 1(-1.05,-2.01,-1.07) 2(-4.20,-4.75,0.53
```

```
[# 176] (0 0 0.07) (0 1 0.89) (1 0 0.89) (1 1 0.14)
 W Out to H (6.17,-6.84,-2.66) H to I 1(-3.42,-3.38,4.80) 2(-5.82,-6.31,1.89)
[# 177] (0 0 0.07) (0 1 0.90) (1 0 0.89) (1 1 0.14)
 W Out to H (6.20,-6.85,-2.67) H to I 1(-3.43,-3.39,4.81) 2(-5.82,-6.31,1.90)
[# 178] (0 0 0.07) (0 1 0.90) (1 0 0.89) (1 1 0.13)
 W Out to H (6.22,-6.87,-2.68) H to I 1(-3.44,-3.40,4.83) 2(-5.83,-6.32,1.91)
[# 179] (0 0 0.07) (0 1 0.90) (1 0 0.90) (1 1 0.13)
 W Out to H (6.24,-6.88,-2.69) H to I 1(-3.45,-3.41,4.85) 2(-5.83,-6.32,1.92)
[# 180] (0 0 0.07) (0 1 0.90) (1 0 0.90) (1 1 0.13)
 W Out to H (6.26,-6.89,-2.71) H to I 1(-3.45,-3.42,4.86) 2(-5.84,-6.32,1.92)
[# 181] (0 0 0.07) (0 1 0.90) (1 0 0.90) (1 1 0.13)
 W Out to H (6.28,-6.91,-2.72) H to I 1(-3.46,-3.43,4.88) 2(-5.84,-6.32,1.93)
[# 182] (0 0 0.07) (0 1 0.90) (1 0 0.90) (1 1 0.13)
 W Out to H (6.30,-6.92,-2.73) H to I 1(-3.47,-3.43,4.89) 2(-5.84,-6.32,1.93)
[# 183] (0 0 0.07) (0 1 0.90) (1 0 0.90) (1 1 0.13)
 W Out to H (6.31,-6.93,-2.74) H to I 1(-3.48,-3.44,4.91) 2(-5.85,-6.33,1.94)
[# 184] (0 0 0.07) (0 1 0.90) (1 0 0.90) (1 1 0.13)
 W Out to H (6.33,-6.94,-2.75) H to I 1(-3.49,-3.45,4.92) 2(-5.85,-6.33,1.95)
[# 185] (0 0 0.07) (0 1 0.90) (1 0 0.90) (1 1 0.12)
 W Out to H (6.35,-6.96,-2.76) H to I 1(-3.49,-3.46,4.94) 2(-5.85,-6.33,1.95)
[# 186] (0 0 0.07) (0 1 0.90) (1 0 0.90) (1 1 0.12)
 W Out to H (6.37,-6.97,-2.77) H to I 1(-3.50,-3.47,4.95) 2(-5.86,-6.33,1.96)
[# 187] (0 0 0.07) (0 1 0.90) (1 0 0.90) (1 1 0.12)
 W Out to H (6.39,-6.98,-2.78) H to I 1(-3.51,-3.48,4.96) 2(-5.86,-6.33,1.96)
[# 188] (0 0 0.07) (0 1 0.90) (1 0 0.90) (1 1 0.12)
 W Out to H (6.40,-6.99,-2.79) H to I 1(-3.51,-3.48,4.98) 2(-5.86,-6.34,1.97)
[# 189] (0 0 0.07) (0 1 0.91) (1 0 0.90) (1 1 0.12)
 W Out to H (6.42,-7.00,-2.80) H to I 1(-3.52,-3.49,4.99) 2(-5.87,-6.34,1.98)
[# 190] (0 0 0.07) (0 1 0.91) (1 0 0.90) (1 1 0.12)
 W Out to H (6.44,-7.01,-2.81) H to I 1(-3.53,-3.50,5.00) 2(-5.87,-6.34,1.98)
[# 191] (0 0 0.07) (0 1 0.91) (1 0 0.91) (1 1 0.12)
 W Out to H (6.46,-7.03,-2.82) H to I 1(-3.53,-3.51,5.02) 2(-5.87,-6.34,1.99)
[# 192] (0 0 0.06) (0 1 0.91) (1 0 0.91) (1 1 0.12)
 W Out to H (6.47,-7.04,-2.83) H to I 1(-3.54,-3.52,5.03) 2(-5.87,-6.34,1.99)
[# 193] (0 0 0.06) (0 1 0.91) (1 0 0.91) (1 1 0.11)
 W Out to H (6.49,-7.05,-2.84) H to I 1(-3.55,-3.52,5.04) 2(-5.88,-6.35,2.00)
[# 194] (0 0 0.06) (0 1 0.91) (1 0 0.91) (1 1 0.11)
 W Out to H (6.50,-7.06,-2.85) H to I 1(-3.55,-3.53,5.06) 2(-5.88,-6.35,2.00)
[# 195] (0 0 0.06) (0 1 0.91) (1 0 0.91) (1 1 0.11)
 W Out to H (6.52,-7.07,-2.86) H to I 1(-3.56,-3.54,5.07) 2(-5.88,-6.35,2.01)
[# 196] (0 0 0.06) (0 1 0.91) (1 0 0.91) (1 1 0.11)
 W Out to H (6.54,-7.08,-2.87) H to I 1(-3.57,-3.54,5.08) 2(-5.89,-6.35,2.01)
[# 197] (0 0 0.06) (0 1 0.91) (1 0 0.91) (1 1 0.11)
 W Out to H (6.55,-7.09,-2.88) H to I 1(-3.57,-3.55,5.09) 2(-5.89,-6.35,2.02)
[# 198] (0 0 0.06) (0 1 0.91) (1 0 0.91) (1 1 0.11)
 W Out to H (6.57,-7.10,-2.89) H to I 1(-3.58,-3.56,5.10) 2(-5.89,-6.35,2.02)
[# 199] (0 0 0.06) (0 1 0.91) (1 0 0.91) (1 1 0.11)
 W Out to H (6.58,-7.11,-2.90) H to I 1(-3.58,-3.56,5.11) 2(-5.89,-6.36,2.03)
*[# 200] (0 0 0.06) (0 1 0.91) (1 0 0.91) (1 1 0.11)
 W Out to H (6.60,-7.12,-2.91) H to I 1(-3.59,-3.57,5.13) 2(-5.90,-6.36,2.03)
```

*See final output values in Fig. 6.B.1 above for weight location by layer: top row of each iteration gives set of values (input 1, input 2, output) for each possible input combination {0,0}; {1,0}; {1,1}.

6.C. Back Propagation Case Study[§]: The XOR Problem — 3 Layer BP Network

The final weights and outputs for XOR *3 layers* Network after 420 iterations are:

$$\begin{array}{llll} \text{input} & \#1 : (0,0) \rightarrow 0.03 & \text{output} & \#1 \\ & \#2 : (0,1) \rightarrow 0.94 & & \#2 \\ & \#3 : (1,0) \rightarrow 0.93 & & \#3 \\ & \#4 : (1,1) \rightarrow 0.07 & & \#4 \end{array}$$

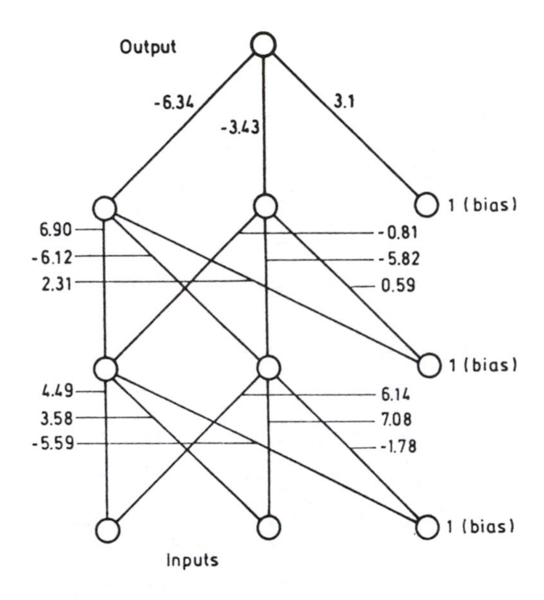

Fig. 6.C.1. Final weight values.

Learning rate: 30 initially, 5 final
Learning rate is reduced by 1 every 10 iterations, namely:
$30, 29, 28, \ldots, 5$
Program: XOR 3.C (C-language)
Purpose: Exclusive-OR function using 3 Hidden Layers

[§]Computed by Sang Lee, EECS Dept., University of Illinois, Chicago, 1993.

Use program of Sec. 6.A up to this point except for denoting XOR 3 instead of XOR 2 where indicated by ← at right-hand side, then, continue here:

```
switch (message) {

    case WM_CREATE:

                /* initialize  weights */
        for (k = 0; k< NOUTPUTS*NHUNITS2; k++)
        {
            weight.weight2[k] = rand()%10;
            if (weight.weight2[k] > 5)
                weight.weight2[k] = -weight.weight2[k];
            weight.weight2[k] = weight.weight2[k]/10;
        }
        for (k = 0; k< NHUNITS2*NHUNITS1; k++)
        {
            weight.weight1[k] = rand()%10;
            if (weight.weight1[k] > 5)
                weight.weight1[k] = -weight.weight1[k];
            weight.weight1[k] = weight.weight1[k]/10;
        }
        for (k = 0; k< NHUNITS1*NINPUTS; k++)
        {
            weight.weight0[k] = rand()%10;
            if (weight.weight0[k] > 5)
                weight.weight0[k] = -weight.weight0[k];
            weight.weight0[k] = weight.weight0[k]/10;
        }

                /* initialize learing rate . this is going to be
                 smaller as time goes on */
        learnrate = 6;
```

XOR3.C
MainWndProc

```
count = 0;
        /* initialize timer */
if(!SetTimer(hWnd,ID_TIMER,500,NULL))
{
    MessageBox(hWnd,"Too many clocks or timers !","Ball",
    MB_ICONEXCLAMATION | MB_OK);
    return FALSE;
}

        /* initialize color pens */
hPenBlack = CreatePenIndirect(&lpBlack);
hPenBlue = CreatePenIndirect(&lpBlue);
hPenYellow = CreatePenIndirect(&lpYellow);
hPenRed = CreatePenIndirect(&lpRed);
hPenGreen = CreatePenIndirect(&lpGreen);
return 0;

case WM_TIMER:

    /* decrease learning rate to have fast convergence */
if ((count%101) == 100)
    learnrate = learnrate/2;

        /****************************************/
        /* put first input into neural network */
        /****************************************/
activ[0] = 0;
activ[1] = 0;
        /* This is bias */
activ[NINPUTS-1] = 1;

        /**********************************/
        /* forward activation propagation */
        /**********************************/

        /* calculate activation for hidden nodes 1 */
for (i = NINPUTS,k=0; i< NINPUTS+NHUNITS1;i++)
{
    net_input[i] = 0;
    for(j = 0; j< NINPUTS;j++,k++)
    {
        net_input[i] += activ[j]*weight.weight0[k];
    }
        /* apply activation function */
    activ[i] = 1/(1+(float)exp(-net_input[i])) ;
}
        /* This is bias */
activ[NINPUTS+NHUNITS1-1] = 1;

        /* calculate activation for hidden nodes 2 */
for (i = NINPUTS+NHUNITS1,k=0; i< NINPUTS+NHUNITS1+NHUNITS2;
 i++)
{
    net_input[i] = 0;
    for(j = NINPUTS; j< NINPUTS+NHUNITS1;j++,k++)
```

```
XOR3.C
              MainWndProc

        {
            net_input[i] += activ[j]*weight.weight1[k];
        }
                    /* apply activation function */
        activ[i] = 1/(1+(float)exp(-net_input[i])) ;
    }
                    /* This is bias */
    activ[NINPUTS+NHUNITS1+NHUNITS2-1] = 1;

                    /* calculate activation for output nodes */
    for (i = NINPUTS+NHUNITS1+NHUNITS2,k=0; i< NINPUTS+NHUNITS1+
     NHUNITS2+NOUTPUTS;i++)
    {
        net_input[i] = 0;
        for(j = NINPUTS+NHUNITS1; j< NINPUTS+NHUNITS1+NHUNITS2;
         j++,k++)
        {
            net_input[i] += activ[j]*weight.weight2[k];
        }
                    /* apply activation function */
        activ[i] = 1/(1+(float)exp(-net_input[i])) ;
    }

                    /* this is final activation of the neural network
                    */
    first = activ[NUNITS-1];

                    /* using target output, calculate weight changes */
                    /* If both inputs are zero, then output should be
                    zero. */
    target[0] = 0;

/* calculate errors for outputs */
    for (i = NUNITS-NOUTPUTS,k=0; i < NUNITS; i++,k++)
    {
        error[i] = target[k] - activ[i];
    }

                    /**************************************************
                    */
                    /* backward error propagation for the first input
                    */
                    /**************************************************
                    */

                    /* initialize errors before doing something else */
    for (i = 0; i < NUNITS-1; i++)
    {
        error[i] = 0;
    }

                    /* calculate errors between outputs and hidden
                    nodes 2 */
    for ( i = NUNITS-NOUTPUTS,k=0: i < NUNITS; i++)
    {
```

XOR3.C
MainWndProc

```
|    |            |      delta[i] = error[i]*activ[i]*(1.0 -activ[i]);
|    |            |                                                  /* error
|    |            |                                      back propagate before
|    |            |                                      bias */
|    |            |      for (j = NUNITS-NOUTPUTS-NHUNITS2; j < NUNITS-NOUTPUTS-1;
|    |            |        j++,k++)
|    |            |           error[j] += delta[i]*weight.weight2[k];
|    |            └──}
|    |
|    |                    /* calculate errors between hidden nodes 2 and
|    |                     hidden nodes 1 */
|    |              for ( i = NUNITS-NOUTPUTS-NHUNITS2,k=0; i < NUNITS-NOUTPUTS-1;
|    |                i++)
|    |            ┌──{
|    |            |      delta[i] = error[i]*activ[i]*(1.0 -activ[i]);
|    |            |                                                  /* error
|    |            |                                      back propaget before
|    |            |                                      bias */
|    |            |      for (j = NUNITS-NOUTPUTS-NHUNITS2-NHUNITS1; j < NUNITS-
|    |            |       NOUTPUTS-NHUNITS2-1; j++,k++)
|    |            |           error[j] += delta[i]*weight.weight1[k];
|    |            └──}
|    |
|    |                    /* calculate errors between hidden nodes 1 and
|    |                     input nodes*/
|    |              for ( i = NUNITS-NOUTPUTS-NHUNITS2-NHUNITS1,k=0; i < NUNITS-
|    |               NOUTPUTS-NHUNITS2-1; i++)
|    |            ┌──{
|    |            |      delta[i] = error[i]*activ[i]*(1.0 -activ[i]);
|    |            |                /* we don't need to calculate errors for input
|    |            |                 nodes */
|    |            └──}
|    |
|    |                    /* calculate delta weight changes between outputs
|    |                     and hidden nodes 2 */
|    |              for ( i = NUNITS-NOUTPUTS,k=0; i < NUNITS; i++)
|    |            ┌──{
|    |            |      for (j = NUNITS-NOUTPUTS-NHUNITS2; j < NUNITS-NOUTPUTS;
|    |            |        j++,k++)
|    |            |           weight.weight2[k] += delta[i]*activ[j]*learnrate;
|    |            └──}
|    |
|    |                    /* calculate delta weight changes between hidden
|    |                     nodes 2 and hidden nodes 1 */
|    |              for ( i = NUNITS-NOUTPUTS-NHUNITS2,k=0; i < NUNITS-NOUTPUTS;
|    |                i++)
|    |            ┌──{
|    |            |      for (j = NUNITS-NOUTPUTS-NHUNITS2-NHUNITS1; j < NUNITS-
|    |            |       NOUTPUTS-NHUNITS2; j++,k++)
|    |            |           weight.weight1[k] += delta[i]*activ[j]*learnrate;
|    |            └──}
|    |
|    |                    /* calculate delta weight changes between hidden
|    |                     1 nodes and input nodes*/
|    |              for ( i = NUNITS-NOUTPUTS-NHUNITS2-NHUNITS1,k=0; i < NUNITS-
```

```
XOR3.C
                MainWndProc

          NOUTPUTS-NHUNITS2; i++)
        {
            for (j = 0; j < NINPUTS; j++,k++)
                weight.weight0[k] += delta[i]*activ[j]*learnrate;
        }
                    /*****************************************/
                    /* put second input into neural network  */
                    /*****************************************/
        activ[0] = 0;
        activ[1] = 1;
                    /* This is bias */
        activ[NINPUTS-1] = 1;

                    /***********************************/
                    /* forward activation propagation */
                    /***********************************/

                    /* calculate activation for hidden nodes 1 */
        for (i = NINPUTS,k=0; i< NINPUTS+NHUNITS1;i++)
        {
            net_input[i] = 0;
            for(j = 0; j< NINPUTS;j++,k++)
            {
                net_input[i] += activ[j]*weight.weight0[k];
            }
                    /* apply activation function */
            activ[i] = 1/(1+(float)exp(-net_input[i])) ;
        }
                    /* This is bias */
        activ[NINPUTS+NHUNITS1-1] = 1;

                    /* calculate activation for hidden nodes 2 */
        for (i = NINPUTS+NHUNITS1,k=0; i< NINPUTS+NHUNITS1+NHUNITS2;
         i++)
        {
            net_input[i] = 0;
            for(j = NINPUTS; j< NINPUTS+NHUNITS1;j++,k++)
            {
                net_input[i] += activ[j]*weight.weight1[k];
            }
                    /* apply activation function */
            activ[i] = 1/(1+(float)exp(-net_input[i])) ;
        }
                    /* This is bias */
        activ[NINPUTS+NHUNITS1+NHUNITS2-1] = 1;

                    /* calculate activation for output nodes */
        for (i = NINPUTS+NHUNITS1+NHUNITS2,k=0; i< NINPUTS+NHUNITS1+
         NHUNITS2+NOUTPUTS;i++)
        {
            net_input[i] = 0;
            for(j = NINPUTS+NHUNITS1; j< NINPUTS+NHUNITS1+NHUNITS2;
             j++,k++)
            {
                net_input[i] += activ[j]*weight.weight2[k];
```

```
XOR3.C
                    MainWndProc

    |    └──}
    |            /* apply activation function */
    |      activ[i] = 1/(1+(float)exp(-net_input[i])) ;
    └──}
                  /* this is final activation of the neural network
                  */
    second = activ[NUNITS-1];

                  /* using target output, calculate weight changes */
                  /* If one of the inputs is not zero, then output
                   should be one */
    target[0] = 1;

/* calculate errors for outputs */
    for (i = NUNITS-NOUTPUTS,k=0; i < NUNITS; i++,k++)
    ┌──{
    |      error[i] = target[k] - activ[i];
    └──}

                  /*************************************************
                   */
                  /* backward error propagation for the second
                   input */
                  /*************************************************/

                  /* initialize errors before doing something else */
    for (i = 0; i < NUNITS-1; i++)
    ┌──{
    |      error[i] = 0;
    └──}

                  /* calculate errors between outputs and hidden
                   nodes 2 */
    for ( i = NUNITS-NOUTPUTS,k=0; i < NUNITS; i++)
    ┌──{
    |      delta[i] = error[i]*activ[i]*(1.0 -activ[i]);
    |                                                   /* error
    |                                        back propagate before
    |                                        bias */
    |      for (j = NUNITS-NOUTPUTS-NHUNITS2; j < NUNITS-NOUTPUTS-1;
    |        j++,k++)
    |          error[j] += delta[i]*weight.weight2[k];
    └──}

                  /* calculate errors between hidden nodes 2 and
                   hidden nodes 1 */
    for ( i = NUNITS-NOUTPUTS-NHUNITS2,k=0; i < NUNITS-NOUTPUTS-1;
      i++)
    ┌──{
    |      delta[i] = error[i]*activ[i]*(1.0 -activ[i]);
    |                                                   /* error
    |                                        back propaget before
    |                                        bias */
    |      for (j = NUNITS-NOUTPUTS-NHUNITS2-NHUNITS1; j < NUNITS-
    |       NOUTPUTS-NHUNITS2-1; j++,k++)
```

```
XOR3.C
                    MainWndProc

                error[j] += delta[i]*weight.weight1[k];
        }

                    /* calculate errors between hidden nodes 1 and
                       input nodes*/
        for ( i = NUNITS-NOUTPUTS-NHUNITS2-NHUNITS1,k=0; i < NUNITS-
          NOUTPUTS-NHUNITS2-1; i++)
        {
            delta[i] = error[i]*activ[i]*(1.0 -activ[i]);
                       /* we don't need to calculate errors for input
                          nodes */
        }

                    /* calculate delta weight changes between outputs
                       and hidden nodes 2 */
        for ( i = NUNITS-NOUTPUTS,k=0; i < NUNITS: i++)
        {
            for (j = NUNITS-NOUTPUTS-NHUNITS2; j < NUNITS-NOUTPUTS;
              j++,k++)
                weight.weight2[k] += delta[i]*activ[j]*learnrate;
        }

                    /* calculate delta weight changes between hidden
                       nodes 2 and hidden nodes 1 */
        for ( i = NUNITS-NOUTPUTS-NHUNITS2,k=0; i < NUNITS-NOUTPUTS;
          i++)
        {
            for (j = NUNITS-NOUTPUTS-NHUNITS2-NHUNITS1; j < NUNITS-
              NOUTPUTS-NHUNITS2; j++,k++)
                weight.weight1[k] += delta[i]*activ[j]*learnrate;
        }

                    /* calculate delta weight changes between hidden
                       1 nodes and input nodes*/
        for ( i = NUNITS-NOUTPUTS-NHUNITS2-NHUNITS1,k=0; i < NUNITS-
          NOUTPUTS-NHUNITS2; i++)
        {
            for (j = 0; j < NINPUTS; j++,k++)
                weight.weight0[k] += delta[i]*activ[j]*learnrate;
        }
                    /***************************************/
                    /* put 3rd input into neural network */
                    /***************************************/

        activ[0] = 1;
        activ[1] = 0;
                    /* This is bias */
        activ[NINPUTS-1] = 1;

                    /************************************/
                    /* forward activation propagation */
                    /************************************/

                    /* calculate activation for hidden nodes 1 */
        for (i = NINPUTS,k=0; i< NINPUTS+NHUNITS1;i++)
```

XOR3.C
MainWndProc

```
|     |         ┌──{
|     |         |       net_input[i] = 0;
|     |         |       for(j = 0; j< NINPUTS;j++,k++)
|     |         |    ┌──{
|     |         |    |       net_input[i] += activ[j]*weight.weight0[k];
|     |         |    └──}
|     |         |              /* apply activation function */
|     |         |       activ[i] = 1/(1+(float)exp(-net_input[i])) ;
|     |         └──}
|     |                     /* This is bias */
|     |            activ[NINPUTS+NHUNITS1-1] = 1;
|     |
|     |                     /* calculate activation for hidden nodes 2 */
|     |            for (i = NINPUTS+NHUNITS1,k=0; i< NINPUTS+NHUNITS1+NHUNITS2;
|     |             i++)
|     |         ┌──{
|     |         |       net_input[i] = 0;
|     |         |       for(j = NINPUTS; j< NUNITS+NHUNITS1;j++,k++)
|     |         |    ┌──{
|     |         |    |       net_input[i] += activ[j]*weight.weight1[k];
|     |         |    └──}
|     |         |              /* apply activation function */
|     |         |       activ[i] = 1/(1+(float)exp(-net_input[i])) ;
|     |         └──}
|     |                     /* This is bias */
|     |            activ[NINPUTS+NHUNITS1+NHUNITS2-1] = 1;
|     |
|     |                     /* calculate activation for output nodes */
|     |            for (i = NINPUTS+NHUNITS1+NHUNITS2,k=0; i< NINPUTS+NHUNITS1+
|     |             NHUNITS2+NOUTPUTS;i++)
|     |         ┌──{
|     |         |       net_input[i] = 0;
|     |         |       for(j = NINPUTS+NHUNITS1; j< NINPUTS+NHUNITS1+NHUNITS2;
|     |         |        j++,k++)
|     |         |    ┌──{
|     |         |    |       net_input[i] += activ[j]*weight.weight2[k];
|     |         |    └──}
|     |         |              /* apply activation function */
|     |         |       activ[i] = 1/(1+(float)exp(-net_input[i])) ;
|     |         └──}
|     |                     /* this is final activation of the neural network
|     |                      */
|     |            third = activ[NUNITS-1];
|     |
|     |                     /* using target output, calculate weight changes */
|     |                     /* If one input is not zero, then output should
|     |                      be one */
|     |            target[0] = 1;
|     |
|     |         /* calculate errors for outputs */
|     |            for (i = NUNITS-NOUTPUTS,k=0; i < NUNITS; i++,k++)
|     |         ┌──{
|     |         |       error[i] = target[k] - activ[i];
|     |         └──}
|     |
```

```
XOR3.C
          MainWndProc

            /****************************************************
            */
            /* backward error propagation for the third input
            */
            /****************************************************/

            /* initialize errors before doing something else */
   for (i = 0; i < NUNITS-1; i++)
   {
       error[i] = 0;
   }

            /* calculate errors between outputs and hidden
            nodes 2 */
   for ( i = NUNITS-NOUTPUTS,k=0; i < NUNITS; i++)
   {
       delta[i] = error[i]*activ[i]*(1.0 -activ[i]);
                                            /* error
                                   back propagate before
                                   bias */
       for (j = NUNITS-NOUTPUTS-NHUNITS2; j < NUNITS-NOUTPUTS-1;
         j++,k++)
           error[j] += delta[i]*weight.weight2[k];
   }

            /* calculate errors between hidden nodes 2 and
            hidden nodes 1 */
   for ( i = NUNITS-NOUTPUTS-NHUNITS2,k=0; i < NUNITS-NOUTPUTS-1;
     i++)
   {
       delta[i] = error[i]*activ[i]*(1.0 -activ[i]);
                                            /* error
                                   back propaget before
                                   bias */
       for (j = NUNITS-NOUTPUTS-NHUNITS2-NHUNITS1; j < NUNITS-
        NOUTPUTS-NHUNITS2-1; j++,k++)
           error[j] += delta[i]*weight.weight1[k];
   }

            /* calculate errors between hidden nodes 1 and
            input nodes*/
   for ( i = NUNITS-NOUTPUTS-NHUNITS2-NHUNITS1,k=0; i < NUNITS-
    NOUTPUTS-NHUNITS2-1; i++)
   {
       delta[i] = error[i]*activ[i]*(1.0 -activ[i]);
                 /* we don't need to calculate errors for input
                 nodes */
   }

            /* calculate delta weight changes between outputs
            and hidden nodes 2 */
   for ( i = NUNITS-NOUTPUTS,k=0; i < NUNITS; i++)
   {
       for (j = NUNITS-NOUTPUTS-NHUNITS2; j < NUNITS-NOUTPUTS;
         j++,k++)
```

```
                    XOR3.C
                              MainWndProc

|   |        |          weight.weight2[k] += delta[i]*activ[j]*learnrate;
|   |        └──}
|   |
|   |                      /* calculate delta weight changes between hidden
|   |                        nodes 2 and hidden nodes 1 */
|   |          for ( i = NUNITS-NOUTPUTS-NHUNITS2,k=0; i < NUNITS-NOUTPUTS;
|   |            i++)
|   |        ┌──{
|   |        |      for (j = NUNITS-NOUTPUTS-NHUNITS2-NHUNITS1; j < NUNITS-
|   |        |        NOUTPUTS-NHUNITS2; j++,k++)
|   |        |          weight.weight1[k] += delta[i]*activ[j]*learnrate;
|   |        └──}
|   |
|   |                      /* calculate delta weight changes between hidden
|   |                        1 nodes and input nodes*/
|   |          for ( i = NUNITS-NOUTPUTS-NHUNITS2-NHUNITS1,k=0; i < NUNITS-
|   |           NOUTPUTS-NHUNITS2: i++)
|   |        ┌──{
|   |        |      for (j = 0; j < NINPUTS; j++,k++)
|   |        |          weight.weight0[k] += delta[i]*activ[j]*learnrate;
|   |        └──}
|   |                      /****************************************/
|   |                      /* put fourth input into neural network */
|   |                      /****************************************/
|   |
|   |          activ[0] = 1;
|   |          activ[1] = 1;
|   |                      /* This is bias */
|   |          activ[NINPUTS-1] = 1:
|   |
|   |                      /**********************************/
|   |                      /* forward activation propagation */
|   |                      /**********************************/
|   |
|   |                      /* calculate activation for hidden nodes 1 */
|   |          for (i = NINPUTS,k=0; i< NINPUTS+NHUNITS1;i++)
|   |        ┌──{
|   |        |      net_input[i] = 0;
|   |        |      for(j = 0; j< NINPUTS;j++)
|   |        |    ┌──{
|   |        |    |      net_input[i] += activ[j]*weight.weight0[k++];
|   |        |    └──}
|   |        |                /* apply activation function */
|   |        |      activ[i] = 1/(1+(float)exp(-net_input[i])) ;
|   |        └──}
|   |                      /* This is bias */
|   |          activ[NINPUTS+NHUNITS1-1] = 1;
|   |
|   |                      /* calculate activation for hidden nodes 2 */
|   |          for (i = NINPUTS+NHUNITS1,k=0; i< NINPUTS+NHUNITS1+NHUNITS2:
|   |           i++)
|   |        ┌──{
|   |        |      net_input[i] = 0;
|   |        |      for(j = NINPUTS; j< NINPUTS+NHUNITS1:j++,k++)
|   |        |    ┌──{
```

XOR3.C
MainWndProc

```
                net_input[i] += activ[j]*weight.weight1[k];
        }
                    /* apply activation function */
            activ[i] = 1/(1+(float)exp(-net_input[i])) ;
    }
                /* This is bias */
    activ[NINPUTS + NHUNITS1 + NHUNITS2-1] = 1;

                /* calculate activation for output nodes */
    for (i = NINPUTS+NHUNITS1+NHUNITS2,k=0; i< NINPUTS+NHUNITS1+
     NHUNITS2+NOUTPUTS;i++)
    {
            net_input[i] = 0;
            for(j = NINPUTS+NHUNITS1; j< NINPUTS+NHUNITS1+NHUNITS2:
             j++,k++)
            {
                net_input[i] += activ[j]*weight.weight2[k];
            }
                    /* apply activation function */
            activ[i] = 1/(1+(float)exp(-net_input[i])) ;
    }

                /* this is final activation of the neural network
                 */
    fourth = activ[NUNITS-1]:

                /* using target output, calculate weight changes */
                /* If both inputs are one, then output should be
                 zero */
    target[0] = 0;

/* calculate errors for outputs */
    for (i = NUNITS-NOUTPUTS,k=0; i < NUNITS; i++,k++)
    {
            error[i] = target[k] - activ[i]:
    }

                /************************************************
                 */
                /* backward error propagation for the fourth
                 input */
                /************************************************/

                /* initialize errors before doing something else */
    for (i = 0; i < NUNITS-1; i++)
    {
            error[i] = 0;
    }

                /* calculate errors between outputs and hidden
                 nodes 2 */
    for ( i = NUNITS-NOUTPUTS,k=0; i < NUNITS; i++)
    {
            delta[i] = error[i]*activ[i]*(1.0 -activ[i]);
                                                    /* error
```

```
                XOR3.C
                        MainWndProc

|    |       |                                      back propagate before
|    |       |                                      bias */
|    |       |    for (j = NUNITS-NOUTPUTS-NHUNITS2; j < NUNITS-NOUTPUTS-1;
|    |       |      j++,k++)
|    |       |        error[j] += delta[i]*weight.weight2[k];
|    |       └──}
|    |
|    |                  /* calculate errors between hidden nodes 2 and
|    |                   hidden nodes 1 */
|    |          for ( i = NUNITS-NOUTPUTS-NHUNITS2,k=0; i < NUNITS-NOUTPUTS-1;
|    |            i++)
|    |       ┌──{
|    |       |    delta[i] = error[i]*activ[i]*(1.0 -activ[i]);
|    |       |                                               /* error
|    |       |                                      back propaget before
|    |       |                                      bias */
|    |       |    for (j = NUNITS-NOUTPUTS-NHUNITS2-NHUNITS1: j < NUNITS-
|    |       |     NOUTPUTS-NHUNITS2-1; j++,k++)
|    |       |        error[j] += delta[i]*weight.weight1[k];
|    |       └──}
|    |
|    |                  /* calculate errors between hidden nodes 1 and
|    |                   input nodes*/
|    |          for ( i = NUNITS-NOUTPUTS-NHUNITS2-NHUNITS1,k=0; i < NUNITS-
|    |           NOUTPUTS-NHUNITS2-1; i++)
|    |       ┌──{
|    |       |    delta[i] = error[i]*activ[i]*(1.0 -activ[i]);
|    |       |                  /* we don't need to calculate errors for input
|    |       |                   nodes */
|    |       └──}
|    |
|    |                  /* calculate delta weight changes between outputs
|    |                   and hidden nodes 2 */
|    |          for ( i = NUNITS-NOUTPUTS,k=0; i < NUNITS: i++)
|    |       ┌──{
|    |       |    for (j = NUNITS-NOUTPUTS-NHUNITS2; j < NUNITS-NOUTPUTS;
|    |       |      j++,k++)
|    |       |        weight.weight2[k] += delta[i]*activ[j]*learnrate;
|    |       └──}
|    |
|    |                  /* calculate delta weight changes between hidden
|    |                   nodes 2 and hidden nodes 1 */
|    |          for ( i = NUNITS-NOUTPUTS-NHUNITS2,k=0; i < NUNITS-NOUTPUTS;
|    |            i++)
|    |       ┌──{
|    |       |    for (j = NUNITS-NOUTPUTS-NHUNITS2-NHUNITS1; j < NUNITS-
|    |       |     NOUTPUTS-NHUNITS2; j++,k++)
|    |       |        weight.weight1[k] += delta[i]*activ[j]*learnrate;
|    |       └──}
|    |
|    |                  /* calculate delta weight changes between hidden
|    |                   1 nodes and input nodes*/
|    |          for ( i = NUNITS-NOUTPUTS-NHUNITS2-NHUNITS1,k=0; i < NUNITS-
|    |           NOUTPUTS-NHUNITS2; i++)
|    |       ┌──{
```

```
XOR3.C
                MainWndProc

|   |        |     for (j = 0; j < NINPUTS; j++,k++)
|   |        |         weight.weight0[k] += delta[i]*activ[j]*learnrate;
|   |        └──}
|   |
|   |                  /* Now save the result in the file */
|   |          hFile = OpenFile("xor3.fil",&OfStruct,OF_EXIST);
|   |          if (hFile >= 0)
|   |        ┌──{
|   |        |     hFile = OpenFile("xor3.fil",&OfStruct,OF_READWRITE);
|   |        |     if (hFile < 0)
|   |        |   ┌──{
|   |        |   └──}
|   |        └──} /* end of exist file */
|   |          else
|   |        ┌──{
|   |        |     hFile = OpenFile("xor3.fil",&OfStruct,OF_CREATE);
|   |        |     if (hFile < 0)
|   |        |   ┌──{
|   |        |   └──}
|   |        └──} /* end of created error file */
|   |          i = 0;
|   |          write(hFile,(char*)&count,2);
|   |          while(i < count)
|   |        ──{
|   |        |                     /* 16 + WEIGHT STRUCT SIZE = 40 */
|   |        |     read(hFile,out,16+sizeof(WEIGHTSTR));
|   |        |     i++;
|   |        └──}
|   |          write(hFile,(char*)&first,4);
|   |          write(hFile,(char*)&second,4);
|   |          write(hFile,(char*)&third,4);
|   |          write(hFile,(char*)&fourth,4);
|   |                       /* write weights */
|   |          write(hFile,(char*)&weight,sizeof(WEIGHTSTR));
|   |          count++;
|   |          close(hFile);
|   |          InvalidateRect(hWnd,NULL,TRUE);
|   |          break;
|   |
|   |      case WM_PAINT:
|   |          GetClientRect(hWnd,&rect);
|   |          hdc = BeginPaint(hWnd,&ps);
|   |          SelectObject(hdc,hPenBlack);
|   |
|   |          wsprintf(out,"Count# %d LearnR %ld.%02ld Blue(0,0:0.%02ld)
|   |           Yellow(0,1:0.%02ld) Red(1,0:0.%02ld) Green(1,1:0.%02ld)",
|   |          count,(long)learnrate,((long)(learnrate*100))%100,
|   |          (((long)(first*100))%100),(((long)(second*100))%100),(((long)(
|   |           third*100))%100),(((long)(fourth*100))%100)
|   |          );
|   |          DrawText(hdc,out,-1,&rect,DT_CENTER);
|   |          Rectangle(hdc,10,10,15,210);
|   |
|   |          SelectObject(hdc,hPenBlue);
|   |          Rectangle(hdc,200, 10+(200 - (int)(((long)(first*100))%100)*2)
```

```
                    XOR3.C
                              MainWndProc

                .210,210);

                SelectObject(hdc,hPenYellow);
                Rectangle(hdc,300, 10+(200 - (int)(((long)(second*100))%100)*
                 2),310,210);

                SelectObject(hdc,hPenRed);
                Rectangle(hdc,400, 10+(200 - (int)(((long)(third*100))%100)*2)
                 ,410,210);

                SelectObject(hdc,hPenGreen);
                Rectangle(hdc,500, 10+(200 - (int)(((long)(fourth*100))%100)*
                 2),510,210);

                EndPaint(hWnd,&ps);
                break;
            case WM_DESTROY:
                PostQuitMessage(0);
                break;

            default:
                return (DefWindowProc(hWnd, message, wParam, lParam));
        }
        return (NULL);
}
```

Samples of Computation Results (3-Layer BP).

iteration
#1

Count#	(0,0:0)	(0,1:1)	(1,0:1)	(1,1:1)	Out1	Out2	Out3	H21	H22	H23	H24	H25	H26	H11	H12	H13	H14	H15	H16
1	0.55	0	0	0.01	-1.76	-1.97	-3.48	-0.09	-1.01	0.2	-0.51	-0.51	0.74	0.11	-0.69	-0.02	0.12	0.51	0.13
2	0.01	0.01	0.01	0.01	-1.62	-1.78	-3.12	-0.16	-1.1	0.06	-0.58	-0.61	0.56	0.12	-0.69	0.01	0.15	0.54	0.19
3	0.01	0.01	0.02	0.05	-1.37	-1.41	-2.32	-0.29	-1.27	-0.22	-0.74	-0.82	0.23	0.16	-0.67	0.07	0.24	0.58	0.31
4	0.04	0.04	0.16	0.88	-0.73	-0.53	-0.66	-0.63	-1.72	-0.91	-1.16	-1.4	-0.64	0.3	-0.62	0.28	0.52	0.68	0.72
5	0.32	0.05	0.2	0.93	-0.59	-0.37	0.64	-0.71	-1.84	-1.05	-1.24	-1.51	-0.78	0.35	-0.59	0.33	0.6	0.72	0.78
6	0.64	0.02	0.04	0.09	-0.78	-0.59	-2.46	-0.66	-1.79	-0.97	-1.21	-1.49	-0.73	0.37	-0.58	0.3	0.62	0.74	0.72
7	0.07	0.06	0.27	0.97	-0.5	-0.28	2.43	-0.77	-1.95	-1.16	-1.3	-1.62	-0.89	0.41	-0.56	0.37	0.69	0.77	0.82
8	0.92	0.57	0.87	0.87	-0.48	-0.24	2.72	-0.77	-1.97	-1.18	-1.31	-1.65	-0.93	0.4	-0.52	0.39	0.68	0.81	0.83
9	0.94	0.73	0.93	0.94	-0.54	-0.3	1.15	-0.76	-1.96	-1.16	-1.3	-1.65	-0.92	0.4	-0.51	0.38	0.68	0.82	0.81
10	0.75	0.04	0.13	0.72	-0.7	-0.51	-3.38	-0.72	-1.91	-1.1	-1.29	-1.63	-0.89	0.39	-0.53	0.36	0.7	0.82	0.79
11	0.03	0.03	0.07	0.29	-0.65	-0.46	-2.58	-0.75	-1.94	-1.14	-1.3	-1.65	-0.92	0.39	-0.53	0.37	0.71	0.82	0.81
12	0.07	0.06	0.23	0.94	-0.5	-0.28	1.26	-0.81	-2.02	-1.24	-1.35	-1.71	-0.99	0.42	-0.52	0.41	0.74	0.84	0.86
13	0.77	0.06	0.24	0.95	-0.5	-0.29	1.61	-0.82	-2.05	-1.26	-1.37	-1.74	-1.02	0.44	-0.51	0.42	0.77	0.85	0.86
14	0.83	0.14	0.76	0.92	-0.53	-0.32	0.48	-0.81	-2.05	-1.26	-1.37	-1.75	-1.03	0.43	-0.48	0.42	0.78	0.88	0.85
15	0.61	0.02	0.04	0.12	-0.65	-0.45	-2.35	-0.78	-2.02	-1.21	-1.35	-1.73	-1	0.44	-0.48	0.4	0.78	0.89	0.81
16	0.08	0.07	0.31	0.97	-0.49	-0.28	2.72	-0.84	-2.11	-1.31	-1.39	-1.8	-1.08	0.46	-0.46	0.44	0.82	0.9	0.85
17	0.94	0.75	0.92	0.93	-0.54	-0.33	0.98	-0.83	-2.09	-1.29	-1.39	-1.79	-1.06	0.46	-0.46	0.43	0.81	0.91	0.84
18	0.72	0.04	0.09	0.48	-0.68	-0.49	-3.52	-0.79	-2.04	-1.22	-1.36	-1.76	-1.02	0.45	-0.47	0.4	0.82	0.91	0.8
19	0.03	0.03	0.06	0.2	-0.63	-0.44	-2.29	-0.81	-2.07	-1.25	-1.37	-1.77	-1.05	0.45	-0.47	0.41	0.83	0.91	0.82
20	0.09	0.07	0.33	0.97	-0.49	-0.28	2.91	-0.86	-2.14	-1.34	-1.42	-1.83	-1.11	0.47	-0.45	0.45	0.86	0.93	0.86
21	0.95	0.82	0.81	0.93	-0.54	-0.35	0.68	-0.85	-2.12	-1.31	-1.4	-1.82	-1.09	0.47	-0.46	0.44	0.85	0.93	0.85
22	0.66	0.03	0.06	0.23	-0.65	-0.46	-2.33	-0.82	-2.09	-1.26	-1.39	-1.8	-1.06	0.47	-0.46	0.42	0.86	0.93	0.81
23	0.09	0.07	0.31	0.97	-0.52	-0.32	2.47	-0.87	-2.16	-1.34	-1.42	-1.85	-1.13	0.49	0.44	0.45	0.89	0.95	0.85
24	0.92	0.63	0.95	0.95	-0.53	-0.33	1.82	-0.86	-2.16	-1.34	-1.42	-1.85	-1.13	0.48	-0.43	0.45	0.89	0.96	0.84
25	0.86	0.24	0.94	0.95	-0.54	-0.33	1.52	-0.86	-2.17	-1.35	-1.42	-1.87	-1.14	0.48	-0.4	0.46	0.88	0.99	0.83
26	0.82	0.13	0.68	0.94	-0.56	-0.35	1.19	-0.86	-2.18	-1.35	-1.42	-1.88	-1.15	0.48	-0.38	0.46	0.89	1.01	0.82
27	0.76	0.06	0.23	0.94	-0.58	-0.38	1.12	-0.86	-2.19	-1.35	-1.43	-1.89	-1.15	0.5	-0.38	0.46	0.91	1.02	0.8
28	0.75	0.05	0.18	0.87	-0.64	-0.44	0.78	-0.85	-2.18	-1.33	-1.42	-1.89	-1.14	0.5	-0.38	0.45	0.93	1.02	0.78
29	0.31	0.07	0.27	0.95	-0.59	-0.39	1.9	-0.88	-2.22	-1.37	-1.44	-1.92	-1.17	0.52	-0.36	0.46	0.96	1.04	0.79
30	0.87	0.28	0.96	0.96	-0.59	-0.38	2.18	-0.88	-2.23	-1.38	-1.45	-1.93	-1.19	0.52	-0.33	0.47	0.96	1.07	0.78
31	0.9	0.45	0.97	0.97	-0.57	-0.36	2.97	-0.88	-2.25	-1.39	-1.45	-1.95	-1.2	0.51	-0.31	0.48	0.96	1.09	0.78
32	0.95	0.84	0.91	0.92	-0.63	-0.41	0.69	-0.86	-2.23	-1.36	-1.44	-1.94	-1.18	0.51	-0.31	0.47	0.95	1.09	0.76
33	0.66	0.03	0.08	0.33	-0.73	-0.52	-2.66	-0.83	-2.19	-1.31	-1.42	-1.91	-1.14	0.51	-0.31	0.44	0.96	1.09	0.72
34	0.06	0.06	0.2	0.88	-0.66	-0.45	-0.41	-0.86	-2.23	-1.36	-1.44	-1.94	-1.18	0.51	-0.31	0.46	0.98	1.1	0.75
35	0.39	0.05	0.15	0.77	-0.71	-0.51	-2.5	-0.84	-2.22	-1.33	-1.43	-1.93	-1.16	0.51	-0.32	0.45	0.99	1.1	0.73
36	0.07	0.07	0.25	0.94	-0.62	-0.42	1.17	-0.88	-2.27	-1.39	-1.46	-1.97	-1.21	0.53	-0.31	0.47	1.01	1.11	0.76

o/p.1 o/p.2 o/p.3 o/p.h ←——— weights ———→

iteration #1	o/p.1	o/p.2	o/p.3	o/p.1	weights layer 3			weights layer 2						weights layer 1					
407	0.04	0.93	0.92	0.08	-6.04	-3.4	2.91	6.78	-6.1	2.27	-0.81	-5.82	0.54	4.54	3.49	-5.92	6.12	7.08	-1.7
408	0.04	0.93	0.92	0.08	-6.07	-3.41	2.93	6.79	-6.1	2.28	-0.81	-5.82	0.54	4.53	3.5	-5.93	6.12	7.08	-1.71
409	0.04	0.94	0.92	0.08	-6.09	-3.41	2.95	6.81	-6.1	2.28	-0.81	-5.82	0.55	4.53	3.51	-5.94	6.13	7.08	-1.72
410	0.03	0.94	0.92	0.08	-6.12	-3.41	2.96	6.82	-6.11	2.28	-0.81	-5.82	0.55	4.52	3.51	-5.94	6.13	7.08	-1.72
411	0.03	0.94	0.92	0.08	-6.14	-3.41	2.98	6.83	-6.11	2.29	-0.81	-5.82	0.56	4.52	3.52	-5.95	6.13	7.08	-1.73
412	0.03	0.94	0.92	0.07	-6.17	-3.41	2.99	6.84	-6.11	2.29	-0.81	-5.82	0.56	4.52	3.53	-5.96	6.13	7.08	-1.74
413	0.03	0.94	0.92	0.07	-6.19	-3.42	3.01	6.84	-6.11	2.29	-0.81	-5.82	0.56	4.51	3.54	-5.96	6.13	7.08	-1.74
414	0.03	0.94	0.92	0.07	-6.21	-3.42	3.02	6.85	-6.11	2.29	-0.81	-5.82	0.57	4.51	3.54	-5.97	6.13	7.08	-1.75
415	0.03	0.94	0.93	0.07	-6.24	-3.42	3.03	6.86	-6.11	2.3	-0.81	-5.82	0.57	4.51	3.55	-5.97	6.13	7.08	-1.75
416	0.03	0.94	0.93	0.07	-6.26	-3.42	3.05	6.87	-6.12	2.3	-0.81	-5.82	0.57	4.5	3.56	-5.98	6.13	7.08	-1.76
417	0.03	0.94	0.93	0.07	-6.28	-3.42	3.06	6.88	-6.12	2.3	-0.81	-5.82	0.58	4.5	3.56	-5.98	6.13	7.08	-1.76
418	0.03	0.94	0.93	0.07	-6.3	-3.43	3.07	6.89	-6.12	2.3	-0.81	-5.82	0.58	4.5	3.57	-5.99	6.13	7.08	-1.77
419	0.03	0.94	0.93	0.07	-6.32	-3.43	3.08	6.9	-6.12	2.3	-0.81	-5.82	0.58	4.49	3.58	-5.99	6.13	7.08	-1.77
420	0.03	0.94	0.93	0.07	-6.34	-3.43	3.1	6.9	-6.12	2.31	-0.81	-5.82	0.59	4.49	3.58	-5.99	6.14	7.08	-1.78

See Fig. 6.B.1 for weight locations by layer output. 1 denoting output for input set 0,0. 2 denoting output for input set 0,1 etc.

Comments:

1. Bias helps to speed up convergence
2. 3 layers are slower than 2 layers
3. Convergence is sudden, not gradual. Also, no relation can be found in this example between rate and convergence speed

Chapter 7

Hopfield Networks

7.1. Introduction

All networks considered until now assumed only forward flow from input to output, namely nonrecurrent interconnections. This guaranteed network stability. Since biological neural networks incorporate *feed-back*, (i.e., they are recurrent), it is natural that certain artificial networks will also incorporate that feature. The Hopfield neural networks [Hopfield, 1982] do indeed employ both feed-forward and feedback. Once feedback is employed, stability cannot be guaranteed in the general case. Consequently, the Hopfield network design must be one that accounts for stability in its settings.

7.2. Binary Hopfield Networks

Figure 7.1 illustrates a recurrent single layer Hopfield network. Though it is basically a single-layer network, its feedback structure makes it effectively to behave as a multi-layer network. The delay in the feedback will be shown to play a major role in its stability. Such a delay is natural to biological neural networks, noting the delay in the synaptic gap and the finite rate of neuronal firing. Whereas Hopfield networks can be of continuous or of binary output, we consider first a binary Hopfield network, to introduce the concepts of the Hopfield network.

The network of Fig. 7.1 thus satisfies

$$z_j = \sum_{i \neq j} w_{ij} y_i(n) + I_j \, ; \quad n = 0, 1, 2 \ldots \tag{7.1}$$

$$y_j(n+1) = \begin{cases} 1 & \forall\ z_j \geq Th_j \\ 0 & \forall\ z_j < Th_j \end{cases} \quad \text{or:} \quad \begin{cases} 1 & \forall\ z_j(n) > Th_j \\ y_j(n) & \forall\ z_j = Th_j \\ 0 & \forall\ z_j < Th_j \end{cases} \tag{7.2}$$

The ii weight in Eq. (7.1) is zero to indicate no self feedback. The 0-state of y becomes -1 in the bipolar case.

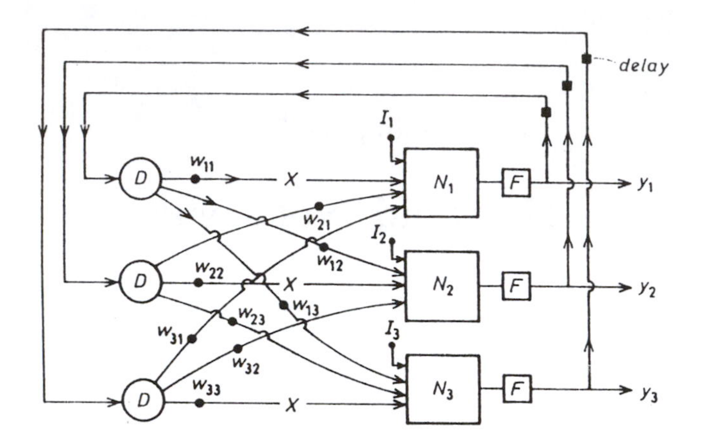

Key : N_i = ith neuron
D = distribution node (external inputs $x_1 \ldots x_3$ also entered to D: not shown)
F = activation function
I_i = bias input
w_{ij} = weight

Fig. 7.1. The structure of a Hopfield network.

By Eqs. (7.1) and (7.2), the Hopfield network employs the basic structure of individual neurons as in the Perceptron or the Adaline. However, by Fig. 7.1, it departs from the previous designs in its feedback structure.

Note that a two neuron binary Hopfield network can be considered as a 2^n state system, with outputs belonging to the four state set {00, 01, 10, 11}. The network, when inputted by an input vector, will stabilize at one of the above states as determined by its weight configurations. A partially incorrect input vector may lead the network to the nearest state to the desired one (to the one related) to the correct input vector).

7.3. Setting of Weights in Hopfield Nets — Bidirectional Associative Memory (BAM) Principle

Hopfield networks employ the principle of BAM (Bidirectional Associative Memory). This implies that the networks' weights are set to satisfy bidirectional associative memory principles; as first proposed by Longuett-Higgins (1968) and also by Cooper (1973) and Kohonen (1977) in relation to other structures, as follows:

Let:

$$\mathbf{x}_i \varepsilon R^m ; \quad \mathbf{y}_i \varepsilon R^n ; \quad i = 1, 2 \cdots L \tag{7.3}$$

and let:

$$W = \sum_i \mathbf{y}_i \mathbf{x}_i^T \tag{7.4}$$

where W is a weight matrix for connections between x and y vector elements. This interconnection is termed as an *associative network*. In particular, when $\mathbf{y}_i = \mathbf{x}_i$ then the connection is termed as *autoassociative*, namely

$$W = \sum_{i=1}^{L} \mathbf{x}_i \mathbf{x}_i^T \quad \text{over} \quad L \text{ vectors} \tag{7.5}$$

such that if the inputs $\mathbf{x}_i$ are orthonormal, namely if

$$\mathbf{x}_i^T \mathbf{x}_j = \delta_{ij} \tag{7.6}$$

then:

$$W\mathbf{x}_i = \mathbf{x}_i \tag{7.7}$$

to retrieve $\mathbf{x}_i$. This setting is called BAM since all $\mathbf{x}_i$ that are associated with the weights W are retrieved whereas the others are not (yielding zero output). Observe that the above implies that W serves as a memory that will allow the network to remember similar input vectors as incorporated in W. The latter structure can be used to reconstruct information, especially incomplete or partly erroneous information.

Specifically, if a single-layer network is considered, then:

$$W = \sum_{i=1}^{L} \mathbf{x}_i \mathbf{x}_i^T \tag{7.8}$$

with:

$$w_{ij} = w_{ji} \quad \forall\, i, j \tag{7.9}$$

by Eq. (7.5)

However, to satisfy the stability requirement to be discussed in Sec. 7.5 below, we also set:

$$w_{ii} = 0 \quad \forall\, i \tag{7.10}$$

to yield the structure of Fig. (7.1) above.

For converting binary inputs $\mathbf{x}(0, 1)$ to yield a *bipolar* (± 1) form, one must set:

$$W = \sum_i (2\mathbf{x}_i - \bar{1})(2\mathbf{x}_i - \bar{1})^T \tag{7.11}$$

$\bar{1}$ being a unify vector, namely

$$\bar{1} \triangleq [1, 1, \ldots, 1]^T \tag{7.12}$$

If certain (or all) inputs are not close to orthogonal, one can first transform them via a Walsh transform as in Sec. 6.4, to yield orthogonal sets for further use.

The BAM feature of the Hopfield NN is what allows it to function with incorrect or partly missing data sets.

Example 7.1:

Let:

$$W = \sum_{i=1}^{L} \mathbf{x}_i \mathbf{x}_i^T = \mathbf{x}_1 \mathbf{x}_1^T + \mathbf{x}_2 \mathbf{x}_2^T + \cdots \mathbf{x}_L \mathbf{x}_L^T$$

with

$$\mathbf{x}_i^T \mathbf{x}_j = \delta_{ij}\,, \quad \mathbf{x}_i = [x_{i1} \cdots x_{in}]^T$$

then, for $n = 2$:

$$W\mathbf{x}_j = \begin{bmatrix} w_{11} & w_{12} \\ w_{21} & w_{22} \end{bmatrix} \begin{bmatrix} x_{j1} \\ x_{j2} \end{bmatrix}$$

such that

$$\begin{aligned} W\mathbf{x}_j &= (\mathbf{x}_1 \mathbf{x}_1^T + \mathbf{x}_2 \mathbf{x}_2^T + \cdots \mathbf{x}_j \mathbf{x}_j^T + \cdots \mathbf{x}_L \mathbf{x}_L^T)\mathbf{x}_j \\ &= \mathbf{x}_1(\mathbf{x}_1^T \mathbf{x}_j) + \mathbf{x}_2(\mathbf{x}_2^T \mathbf{x}_j) + \cdots \mathbf{x}_j(\mathbf{x}_j^T \mathbf{x}_j) + \cdots \mathbf{x}_L(\mathbf{x}_L^T \mathbf{x}_j) = \mathbf{x}_j(\mathbf{x}_j^T \mathbf{x}_j) = \mathbf{x}_j \end{aligned}$$

as long as the inputs are orthonormal.

The degree of closeness of a pattern (input vector) to a memory is evaluated by the *Hamming Distance* [Hamming, 1950, see also: Carlson, 1986, p. 473]. The number of terms in which an error exists in a network (regardless of magnitude of error) is defined as the Hamming distance for that network to provide a measure of distance between an input and the memory considered.

Example 7.2:

Let $\mathbf{x}_i$ be given as a 10-dimensional vectors $\mathbf{x}_1$ and $\mathbf{x}_2$, such that

$$\mathbf{x}_1^T \mathbf{x}_2 = [1 \ \ -1 \ \ -1 \ \ 1 \ \ -1 \ \ 1 \ \ 1 \ \ -1 \ \ -1 \ \ 1] \begin{bmatrix} 1 \\ 1 \\ 1 \\ -1 \\ -1 \\ -1 \\ 1 \\ 1 \\ -1 \\ -1 \end{bmatrix} = -2$$

In that case, the Hamming distance d is:

$$d(\mathbf{x}_1, ; \mathbf{x}_2) = 6$$

while

$$\mathbf{x}_i^T \mathbf{x}_i = \dim(\mathbf{x}_i) = 10 \qquad \text{for } i = 1, 2$$

such that

$$d = \frac{1}{2} \left[\dim(\mathbf{x}) - \mathbf{x}_1^T \mathbf{x}_2\right]$$

Hence the net will emphasize an input that (nearly) *belongs* to a given training set and *de-emphasize* those inputs that do not (nearly) belong (is associated — hence the term "BAM").

7.4. Walsh Functions

Walsh functions were proposed by J. L. Walsh in 1923 (see Beauchamp, 1984). They form an ordered set of rectangular (stair-case) values +1, −1 defined over a limited time interval t. The Walsh function WAL(n, t) is thus defined by an ordering number n and the time period t, such that:

$$x(t) = \sum_{i=0}^{N-1} X_i Wal(i, t) \tag{7.13}$$

Walsh functions are orthogonal, s.t.

$$\sum_{t=0}^{N-1} Wal(m, t) Wal(n, t) = \begin{cases} N & \forall\, n = m \\ 0 & \forall\, n \neq m \end{cases} \tag{7.14}$$

Consider a time series $\{x_i\}$ of N samples: The Walsh Transform (WT) of $\{x_i\}$ is given by X_n where

$$X_n = \frac{1}{N} \sum_{i=0}^{N-1} x_i Wal(n, i) \tag{7.15}$$

and the IWT (inverse Walsh transform) is:

$$x_i = \sum_{n=0}^{N-1} X_n Wal(n, i) \tag{7.16}$$

where

$$i, n = 0, 1, \ldots, N-1 \tag{7.17}$$

X_n thus being the discrete Walsh transform and x_i being its inverse, in parallelity to the discrete Fourier Transform of x_i, which is given by:

$$X_k = \sum_{n=0}^{N-1} x_n F_N^{nk} \tag{7.18}$$

and to the IFT (inverse Fourier transform), namely:

$$x_n = \frac{1}{N} \sum_{k=0}^{N-1} X_k F_n^{-nk} \tag{7.19}$$

where:

$$F_N = \exp(-j2\pi/N) \tag{7.20}$$

Hence, to apply BAM to memories (vectors) that are not orthonormal we may first transform them to obtain their orthogonal Walsh transforms and then apply BAM to these transforms.

Example 7.4:

Table 7.1. (*Reference:* K. Beauchamp, Sequence and Series, *Encyclopedia of Science and Technology,* Vol. 12, pp. 534–544, 1987. Courtesy of Academic Press, Orlando, FL.

i, t	Wal (i, t)							
0, 8	1	1	·	·	·	·	·	1
1, 8	1	1	1	1	−1	−1	−1	−1
2, 8	1	1	−1	−1	−1	−1	1	1
3, 8	1	1	−1	−1	1	1	−1	−1
4, 8	1	−1	−1	1	1	−1	−1	1
5, 8	1	−1	−1	1	−1	1	1	−1
6, 8	1	−1	1	−1	−1	1	−1	1
7, 8	1	−1	1	−1	1	−1	1	−1

7.5. Network Stability

Weight adjustment in a feedback network must guarantee network stability. It was shown by Cohen and Grossberg (1983) that recurrent networks can be guaranteed to be stable if the W matrix of weights is symmetrical and if its diagonal is zero, namely

$$w_{ij} = w_{ji} \quad \forall\ i, j \tag{7.21}$$

with

$$w_{ii} = 0 \quad \forall\ i \tag{7.22}$$

The above requirements result from the Lyapunov stability theorem which states that a system (network) is stable if an energy function (its Lyapunov function) can be defined for that system which is guaranteed to always decrease over time [Lyapunov, 1907, see also Sage and White, 1977].

Network (or system) stability can be satisfied via the Lyapunov stability theorem if a function E of the states y of the network (system) can be defined, that satisfies the following conditions:

Condition (A): Any finite change in the states y of the network (system) yields a finite decrease in E.

Condition (B): E is bounded from below.

We thus define an energy function E (denoted also as Lyapunov function) as

$$E = \sum_j Th_j y_j - \sum_j I_j y_j - \frac{1}{2} \sum_i \sum_{j \neq i} w_{ij} y_j y_i \tag{7.23}$$

i denoting the ith neuron
j denoting the jth neuron
I_j being on external input to neuron j
Th_j being the threshold to neuron j

w_{ij} being an element of the weight matrix W, to denote the weight from the output of neuron i to the input of neuron j.

We now prove the network's stability by the Lyapunov theorem as follows: First we set W to be symmetric with all diagonal elements being zero, namely

$$W = W^T \tag{7.24a}$$

$$w_{ii} = 0 \quad \forall\, i \tag{7.24b}$$

and where $|w_{ij..}|$ are bounded for all i, j.

We prove that E satisfies condition (A) of the Lyapunov stability theorem, by considering a change in and only in *one* component $y_k(n+1)$ of the output layer: Denoting $E(n)$ as E at the nth iteration and $y_k(n)$ as y_k at that same iteration, we write:

$$\begin{aligned} \Delta E_n &= E(n+1) - E(n) \\ &= [y_k(n) - y_k(n+1)] \cdot \left[\sum_{i \neq k} w_{ik} y_i(n) + I_k - Th_k \right] \end{aligned} \tag{7.25}$$

We observe via Eq. (7.2) that the binary Hopfield neural network must satisfy that

$$y_k(n+1) = \begin{cases} 1 & \cdot\cdot \quad \forall\, z_k(n) > Th_k \\ y_k(n) & \cdots \quad \forall\, z_k(n) = Th_k \\ 0 & \cdots \quad \forall\, z_k(n) < Th_k \end{cases} \tag{7.26}$$

where

$$z_k = \sum_i w_{ik} y_i + I_k \tag{7.27}$$

and where Th_k denotes the threshold to the given (kth) neuron. Therefore, y_k can undertake only two changes in value:

(i)

$$\text{If} \quad y_k(n) = 1 \quad \text{then } y_k(n+1) = 0 \tag{7.28}$$

(ii)

$$\text{If} \quad y_k(n) = 0 \quad \text{then } y_k(n+1) = 1 \tag{7.29}$$

Now, under scenario (i):

$$[y_k(n) - y_k(n+1)] > 0 \tag{7.30}$$

However, this can occur only if

$$\sum_{i \neq k} w_{ik} y_i + I_k - Th_k = z_k(n) - Th_k < 0 \tag{7.31}$$

by Eq. (7.26) above. Hence, by Eq. (7.25) $\Delta E < 0$, such that E is reduced as required by condition (A) of the Lyapunov Stability theorem.

Similarly under scenario (ii);

$$[y_k(n) - y_k(n+1)] < 0 \tag{7.32}$$

However, this can occur only if

$$\sum_i w_{ik} y_i + I_k - Th_k = z_k(n) - Th_k > 0 \tag{7.33}$$

by Eq. (7.26) above. Hence, again $\Delta E < 0$ such that E is again reduced as required.

Finally, condition (B) of the Lyapunov stability theorem is trivially satisfied since in the worst case (the most negative-energy case) all $y_i = y_j = 1$ such that

$$E = -\sum_i \sum_j |w_{ij}| - \sum_i |I_i| + \sum_i Th_i \tag{7.34}$$

which is bounded from below noting that w_{ij} must all be finite and bounded.

The proof also holds for situations where several y_j terms are changed. Also, note that in the feedback interconnection of the Hopfield network we have that:

$$z_i = \sum_{i \neq k} w_{ik} y_j \tag{7.35}$$

However, if $w_{ii} \neq 0$, then ΔE as in Eq. (7.25) would have included terms of the form of

$$-w_{kk} y_k^2(n+1) \Delta y$$

which could be either positive or negative, depending on the sign of w_{ii} and on the old and new values of y_i. This would have violated the convergence proof above.

Lack of symmetry in the W matrix would invalidate the expressions used in the present proof.

7.6. Summary of the Procedure for Implementing the Hopfield Network

Let the weight matrix of the Hopfield network satisfy:

$$W = \sum_{i=1}^{L}(2\mathbf{x}_i - \bar{1})(2\mathbf{x}_i - \bar{1})^T \quad (1)^*$$

$L =$ numbers of training sets:

The computation of the Hopfield network with BAM memory as in Eq. (1)* above, will then proceed according to the following *procedure*.

(1) Assign weights w_{ij} of matrix W according to Eq. (1)*, with $w_{ii} = 0 \; \forall \, i$ and $\boldsymbol{x}_i$ being the training vectors of the network.

(2) Enter an unknown input pattern $\mathbf{x}$ and set:

$$y_i(0) = x_i \quad (2)^*$$

where x_i is the ith element of vector $\mathbf{x}$ considered.

(3) Subsequently, iterate:

$$y_i(n+1) = f_N\left[z_i(n)\right] \quad (3)^*$$

where f_n denotes the activation function,

$$f_N(z) = \begin{cases} 1 \cdots \forall \; z > Th \\ \text{unchanged} \; \cdots \forall \; z = Th \\ -1 \cdots \forall \; z < Th \end{cases} \quad (4)^*$$

and where

$$z_i(n) = \sum_{i=1} w_{ij} y_i(n) \quad (5)^*$$

n being all integer denoting the number of the iteration $(n = 0, 1, 2, \ldots)$.

Continue iteration until convergence is reached, namely, until changes in $y_i(n+1)$ as compared with $y_i(n)$ are below some low threshold value.

(4) The process is repeated for all elements of the unknown vector above by going back to Step (2) while choosing the next element until all elements of the vector have been so processed.

(5) As long as new (unknown) input vectors exist for a given problem, go to the next input vector $\mathbf{x}$ and return to Step (2) above.

The node outputs to which $y(n)$ converge per each unknown input vector $\mathbf{x}$ represent the exemplar (training) vector which best represents (best matches) the unknown input.

For each element of input of the Hopfield network there is an output. Hence, for a character recognition problem in a 5×5 grid there are 25 inputs and 25 outputs.

7.7. Continuous Hopfield Models

The discrete Hopfield network has been extended by Hopfield and others to a continuous form as follows: Letting z_i be the net summed output, then the network output y_i satisfies

$$y_i = f_i(\lambda z_i) = \frac{1}{2}\left[1 + \tanh(\lambda z_i)\right] \qquad (7.36)$$

as in Fig. 7.2. Note that λ determines the slope of f at $y = \frac{1}{2}$, namely the rate of rise of y.

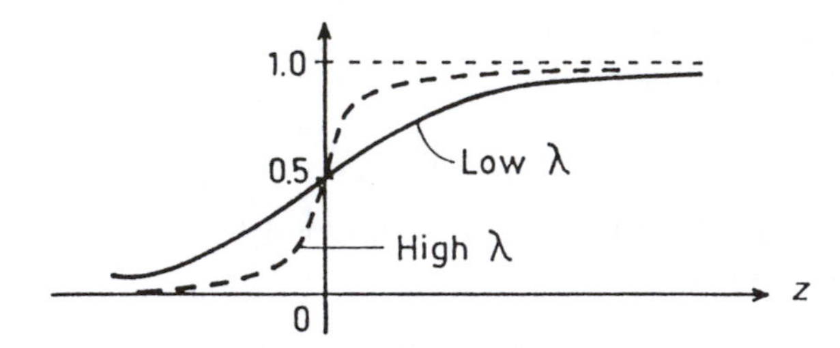

Fig. 7.2. Activation function with variable-λ.

Also, a differential equation can replace the time delay relation between input and network summed output. Hence, the steady state model for the circuit of Fig. 7.3:

$$\sum_{j \neq i} T_{ij} y_j - \frac{z_i}{R_i} + I_i = 0 \qquad (7.37)$$

satisfies the transient equation

$$C\frac{dz_i}{dt} = \sum_{j \neq i} T_{ij} y_j - \frac{z_i}{R_i} + I_i \qquad (7.38a)$$

where

$$y_i = f_N(z_i) \qquad (7.38b)$$

as in Fig. 7.3:

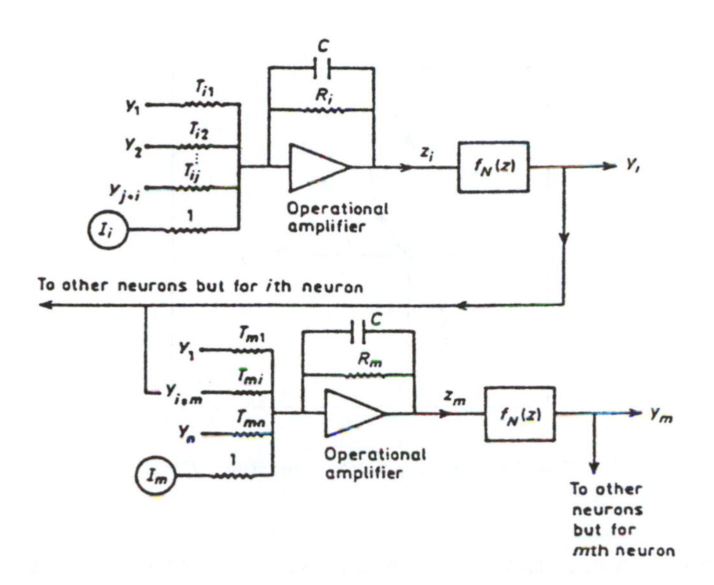

Fig. 7.3. A continuous Hopfield net.

7.8. The Continuous Energy (Lyapunov) Function

Consider the continuous energy function E where:

$$E = -\frac{1}{2}\sum_i \sum_{j \neq i} T_{ij} y_i y_j + \frac{1}{\lambda}\int_0^{y_i} f^{-1}(y)dy - \sum_i I_i y_i \tag{7.39}$$

E above yields that, noting Eq. (7.36):

$$\frac{dE}{dt} = -\sum_i \frac{dy_i}{dt}\left(\sum_{j \neq i} T_{ij} y_j - \frac{z_i}{R_i} + I_i\right) = -\sum_i C \frac{dy_i}{dt}\frac{dz_i}{dt} \tag{7.40}$$

the last equality being due to Eq. (7.38).

Since $z_i = f^{-1}(y_i)$, we can write

$$\frac{dz_i}{dt} = \frac{df^{-1}(y_i)}{dy_i}\frac{dy_i}{dt} \tag{7.41}$$

to yield, via Eq. (7.40) that:

$$\frac{dE}{dt} = -C\frac{df^{-1}(y_i)}{dy_i}\left(\frac{dy_i}{dt}\right)^2 \tag{7.42}$$

Since $f^{-1}(v)$ monotonously increases with v, as in Fig. 7.4, $\frac{dE}{dt}$ always decreases; *to satisfy the Lyapunov stability criterion* as earlier stated. Note that the minimum of E exists by the similarity of E to the energy function of the bipolar case and noting the limiting effect of $f(v)$, as in Fig. 7.2.

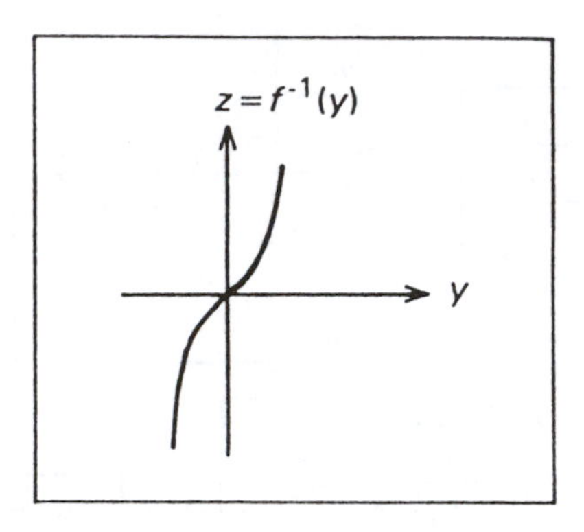

Fig. 7.4. Inverse activation function: f^{-1}.

One important application of the continuous Hopfield network that is worth mentioning is to the Traveling-Salesman problem (TSP) and to related NP-Complete problems (see: Hopfield, J. J. and Tank, D. W., *Biol. Cybern.* **5**, 141–152, 1985). In these problems, the Hopfield network yields extremely fast solutions which, if not optimal (for, say, a high number of cities in the Traveling-Salesman problem) are within a reasonable percentage error of the optimum value, after a small number of iterations. These should compare to the theoretically needed calculations of the order of $(N-1)!$ For N cities for a truly optimal solution. This illustrates a very important property of neural networks in general. They yield a good working solution in reasonable time (number of iterations) for many problems that are very complex and which otherwise may often defy any exact solution. Even though the Hopfield network may not be the best neural network for many of these problems, especially those that defy any analytical description (networks such as Back Propagation of the LAMSTAR may be often the way to go in many such cases, and can also be applied to the TSP problem), for good approximations of NP-complete problems the Hopfield network is the way to go. In these cases, and where the exact solution is often available, one can also compute the error of the network relative to the exact solution. When a problem defies any analysis, this is, of course, not possible. Appendix 7.B below presents a case of applying the Hopfield NN to the TSP problem (with computed results for up to 25 cities).

7.A. Hopfield Network Case Study*: Character Recognition

7.A.1. *Introduction*

The goal of this case study is to recognize three digits of '0', '1', '2' and '4'. To this end, a one-layer Hopfield network is created, it is trained with standard data sets (8*8); make the algorithm converge and it is tested the network with a set of test data having 1, 3, 5, 10, 20, 30-bit errors.

7.A.2. *Network design*

The general Hopfield structure is given in Fig. 7.A.1:

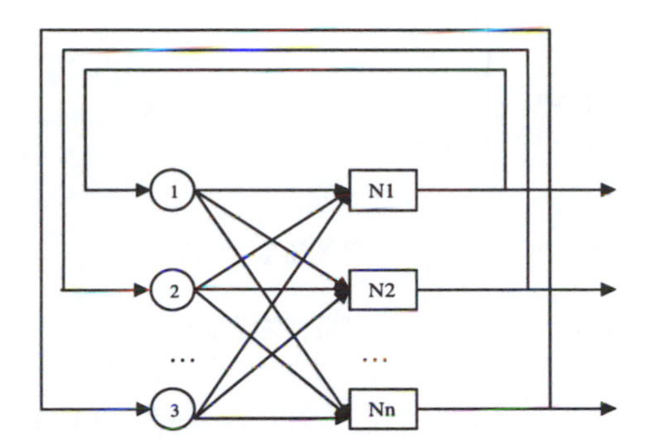

Fig.7.A.1. Hopfield network: a schematic diagram.

The Hopfield neural network is designed and applied to the present case study (using MATLAB) to create a default network:

Example 1: Creating a 64-neuron Hopfield network with initial random weights

```
%% Example #1:
%%      neuronNumber = 64
%%      weitghtCat = 'rand'
%%      defaultWeight = [−5 5]
%% use:
%%      hopfield = createDefaultHopfield(neuronNumber, 'rand', [−5 5])
```

Example 2: Creating a 36-neuron Hopfield network with initial weights of 0.5

```
%%      neuronNumber = 36
%%      weitghtCat = 'const'
%%      defaultWeight = '0.5'
%% use:
%%      hopfield = createDefaultHopfield(neuronNumber, 'const', 0.5)
```

*Computed by Sang K. Lee, EECS Dept., University of Illinois, Chicago, 1993.

7.A.3. *Setting of weights*

(a) The training data set

The training data set applied to the Hopfield network is illustrated as follows:

```
%%%%%%%%%%%%%%%%%%%%%%%%%%%%%%
%% 2
%%%%%%%%%%%%%%%%%%%%%%%%%%%%%%
trainingData(1).input = [...
    -1 -1 -1 -1 -1 -1 -1 -1;...
    -1 -1 -1 -1 -1 -1 -1 -1;...
     1  1  1  1  1 -1 -1  1;...
     1  1  1  1 -1 -1  1  1;...
     1  1  1 -1 -1  1  1  1;...
     1  1 -1 -1  1  1  1  1;...
     1 -1 -1 -1 -1 -1 -1 -1;...
     1 -1 -1 -1 -1 -1 -1 -1 ...
  ];
trainingData(1).name = '2';
%%%%%%%%%%%%%%%%%%%%%%%%%%%%%%
%% 1
%%%%%%%%%%%%%%%%%%%%%%%%%%%%%%
trainingData(2).input = [...
    1 1 1 -1 -1 1 1 1;...
    1 1 1 -1 -1 1 1 1;...
    1 1 1 -1 -1 1 1 1;...
    1 1 1 -1 -1 1 1 1;...
    1 1 1 -1 -1 1 1 1;...
    1 1 1 -1 -1 1 1 1;...
    1 1 1 -1 -1 1 1 1;...
    1 1 1 -1 -1 1 1 1 ...
  ];
trainingData(2).name = '1';
%%%%%%%%%%%%%%%%%%%%%%%%%%%%%%
%% 4
%%%%%%%%%%%%%%%%%%%%%%%%%%%%%%
trainingData(3).input = [...
    -1 -1  1  1  1  1  1  1;...
    -1 -1  1  1  1  1  1  1;...
    -1 -1  1  1 -1 -1  1  1;...
    -1 -1  1  1 -1 -1  1  1;...
    -1 -1 -1 -1 -1 -1 -1 -1;...
    -1 -1 -1 -1 -1 -1 -1 -1;...
     1  1  1  1 -1 -1  1  1;...
     1  1  1  1 -1 -1  1  1 ...
  ];
```

```
trainingData(3).name = '4';
%%%%%%%%%%%%%%%%%%%%%%%%%%%%%%
%% 0
%%%%%%%%%%%%%%%%%%%%%%%%%%%%%%
trainingData(4).input = [...
    1 -1 -1 -1 -1 -1 -1  1;...
    1 -1 -1 -1 -1 -1 -1  1;...
    1 -1 -1  1  1 -1 -1  1;...
    1 -1 -1  1  1 -1 -1  1;...
    1 -1 -1  1  1 -1 -1 -1;...
    1 -1 -1  1  1 -1 -1 -1;...
    1 -1 -1 -1 -1 -1 -1  1;...
    1 -1 -1 -1 -1 -1 -1  1 ...
  ];
trainingData(4).name = '0';
```

(b) Initial Weights:

(1) Get all training data vectors X_i, $i = 1, 2 \ldots L$
(2) Compute the weight matrix $W = \Sigma X_i X_i^T$ over L vectors
(3) Set $w_{ii} = 0$, for all i, where w_{ii} is the ith diagonal element of the weight matrix
(4) Assign the jth row vector of the weight matrix to the j-th neuron as its initial weights.

7.A.4. *Testing*

The test data set is generated by a procedure which adds a specified number of error bits to the original training data set. In this case study, a random procedure is used to implement this function.

Example:

testingData = getHopfieldTestingData(trainingData, numberOfBitError, numberPerTrainingSet)

where the parameter, 'numberOfBitError', is to specify the expected number of bit errors; 'numberPerTrainingSet' is to specify the expected size of the testing data set. The expected testing data set is obtained via the output parameter 'testingData'.

7.A.5. *Results and conclusions*

(a) Success rate VS bit errors

In this experiment, a 64-neuron 1-layer Hopfield is used. The success rate is tabulated as follows:

Success Rate		Number of Testing data set		
		12	100	1000
Number of Bit Error	1	100%	100%	100%
	3	100%	100%	100%
	5	100%	100%	100%
	10	100%	100%	100%
	15	100%	100%	100%
	20	100%	100%	99.9%
	25	100%	98%	97.3%
	30	83.3333%	94%	94.2%
	35	91.6667%	93%	88.8%
	40	83.3333%	82%	83.6%

(b) Conclusion

(1) The Hopfield network is robust with high convergence rate
(2) Hopfield network has high success rate even if in the case of large bit errors.

7.A.6. *MATALAB codes*

File #1

```
function hopfield = nnHopfield

\%\% Create a default Hopfield network
hopfield = createDefaultHopfield(64, ‘const’, 0);

\%\% Training the Hopfield network
trainingData = getHopfieldTrainingData;
[hopfield] = trainingHopfield(hopfield, trainingData);

\%\% test the original training data set;
str = [];
tdSize = size(trainingData);
for n = 1: tdSize(2);
    [hopfield, output] = propagatingHopfield(hopfield,
     trainingData(n).input, 0, 20);

    [outputName, outputVector, outputError] =
     hopfieldClassifier(hopfield, trainingData);

    if strcmp(outputName, trainingData(n).name)
        astr = [num2str(n), ‘==> Succeed!! The Error Is:’,
        num2str(outputError)];
```

```
    else
        astr = [num2str(n), '==> Failed!!'];
    end

    str = strvcat(str, astr);
end
output = str;
display(output);

\%\% test on the testing data set with bit errors
testingData = getHopfieldTestingData(trainingData, 4, 33);
trdSize = size(trainingData);
tedSize = size(testingData);
str = [];
successNum = 0;
for n = 1: tedSize(2)
    [hopfield, output, nInterationNum] = propagatingHopfield(hopfield,
testingData(n).input, 0, 20);
    [outputName, outputVector, outputError] = hopfieldClassifier(hopfield,
trainingData);

    strFormat = '      ';
    vStr = strvcat(strFormat,num2str(n),num2str(nInterationNum));
    if strcmp(outputName, testingData(n).name)
        successNum = successNum + 1;
        astr = [vStr(2,:), '==> Succeed!! Iternation # Is:,', vStr(3,:),
'The Error Is:', num2str(outputError)];
    else
        astr = [vStr(2,:), '==> Failed!! Iternation # Is:,', vStr(3,:),];
    end

    str = strvcat(str, astr);
end

astr = ['The success rate is: ', num2str(successNum * 100/ tedSize(2)),'\%'];
str = strvcat(str, astr);
testResults = str;
display(testResults);
```

File #2

```
function [hopfield, output, nInterationNum] = propagatingHopfield(hopfield,
inputData, errorThreshold, interationNumber)

output = [];
if nargin < 2
   display('propagatingHopfield.m needs at least two parameter');
   return;
```

```
end

if nargin == 2
   errorThreshold = 1e-7;
   interationNumber = [];
end

if nargin == 3
   interationNumber = [];
end

% get inputs
nnInputs = inputData(:)';
nInterationNum = 0;
dError = 2* errorThreshold + 1;
while dError > errorThreshold
  nInterationNum = nInterationNum + 1;
  if ~isempty(interationNumber)
       if nInterationNum > interationNumber
          break;
       end
    end

%% interation here
dError = 0;
output = [];
analogOutput= [];
for ele = 1:hopfield.number
   % retrieve one neuron
   aNeuron = hopfield.neurons(ele);

   % get analog outputs
   z = aNeuron.weights * nnInputs';
   aNeuron.z = z;

   analogOutput = [analogOutput, z];

   % get output
   Th = 0;
   if z > Th
      y = 1;
   elseif z < Th
      y = -1;
   else
      y = z;
   end

   aNeuron.y = y;

   output = [output, aNeuron.y];
```

```
   % update the structure
   hopfield.neurons(ele) = aNeuron;

   % get the error
   newError = (y - nnInputs(ele)) * (y - nnInputs(ele));
   dError = dError + newError;
end

hopfield.output = output;
hopfield.analogOutput = analogOutput;
hopfield.error = dError;

%% feedback
   nnInputs = output;
   %% for tracing only
   %nInterationNum, dError
end

return;
```

File #3

```
function hopfield = trainingHopfield(hopfield, trainingData )

if nargin < 2
   display('trainingHopfield.m needs at least two parameter');
   return;
end

datasetSize = size(trainingData);
weights = [];
for datasetIndex = 1: datasetSize(2)
   mIn = trainingData(datasetIndex).input(:);

   if isempty(weights)
      weights = mIn * mIn';
   else
      weights = weights + mIn * mIn';
   end
end

wSize = size(weights);
for wInd = 1: wSize(1)
   weights(wInd, wInd) = 0;
   hopfield.neurons(wInd).weights = weights(wInd,:);
end

hopfield.weights = weights;
```

File #4

```
function [outputName, outputVector, outputError] = hopfieldClassifier(hopfield,
trainingData)

outputName = [];
outputVector = [];

if nargin < 2
   display('hopfieldClassifier.m needs at least two parameter');
   return;
end

dError = [];
dataSize = size(trainingData);
output = hopfield.output';
for dataInd = 1 : dataSize(2)
   aSet = trainingData(dataInd).input(:);

   vDiff = abs(aSet - output);
   vDiff = vDiff.^2;

   newError = sum(vDiff);

   dError = [dError, newError];
end

if ~isempty(dError)
   [eMin, eInd] = min(dError);
   outputName = trainingData(eInd).name;
   outputVector = trainingData(eInd).input;
   outputError = eMin;
end
```

File #5

```
%%%%%%%%%%%%%%%%%%%%%%%%%%%%%%%%%%%%%%%%%%%%%%%%%%%%%%%%%%%%%%%%%%%%%%%%%%%%%%%%%%
%% A function to create a default one layer Hopfield model
%%
%% input parameters:
%% neuronNumber, to specify all neuron number
%%
%% defaultWeight, to set the default weights
%%
%% Example #1:
%%     neuronNumber = 64
%%     weitghtCat = 'rand'
%%     defaultWeight = [-5 5]
%%  use:
%%     hopfield = createDefaultHopfield(neuronNumber, 'rand', [-5 5])
```

```
%%
%% Example #2:
%%     neuronNumber = 36
%%     weitghtCat = 'const'
%%     defaultWeight = '0.5'
%%  use:
%%     hopfield = createDefaultHopfield(neuronNumber, 'const', 0.5)
%%
%% Author: Yunde Zhong
%%%%%%%%%%%%%%%%%%%%%%%%%%%%%%%%%%%%%%%%%%%%%%%%%%%%%%%%%%%%%%%%%%%%%%%%%%%%%%%%%%%
function hopfield = createDefaultHopfield(neuronNumber, weightCat,
defaultWeight)

hopfield = [];

if nargin < 3
   display('createDefaultHopfield.m needs at least two parameter');
   return;
end

aLayer.number = neuronNumber;
aLayer.error = [];
aLayer.output = [];
aLayer.neurons = [];
aLayer.analogOutput = [];
aLayer.weights = [];

%% create a default layer
for ind = 1: aLayer.number
   %% create a default neuron
   inputsNumber = neuronNumber;
   if strcmp(weightCat, 'rand')
      offset = (defaultWeight(1) + defaultWeight(2))/2.0;
      range = abs(defaultWeight(2) - defaultWeight(1));
      weights = (rand(1,inputsNumber) -0.5 )* range + offset;
   elseif strcmp(weightCat, 'const')
      weights = ones(1,inputsNumber) * defaultWeight;
   else
      error('error paramters when calling createDefaultHopfield.m');
      return;
   end
   aNeuron.weights = weights;
   aNeuron.z = 0;
   aNeuron.y = 0;

   aLayer.neurons = [aLayer.neurons, aNeuron];
   aLayer.weights = [aLayer.weights; weights];
end

hopfield = aLayer;
```

File #6

```
function testingData = getHopfieldTestingData(trainingData, numberOfBitError,
numberPerTrainingSet)

testingData = [];

tdSize = size(trainingData);
tdSize = tdSize(2);

ind = 1;
for tdIndex = 1: tdSize
   input = trainingData(tdIndex).input;
   name = trainingData(tdIndex).name;
   inputSize = size(input);

   for ii = 1: numberPerTrainingSet
       rowInd = [];
       colInd = [];

       flag = ones(size(input));
       bitErrorNum = 0;
       while bitErrorNum < numberOfBitError
          x = ceil(rand(1) * inputSize(1));
          y = ceil(rand(1) * inputSize(2));
          if x <= 0
             x = 1;
          end
          if y <= 0
             y = 1;
          end
          if flag(x, y) ~= -1
             bitErrorNum = bitErrorNum + 1;
             flag(x, y) == -1;
             rowInd = [rowInd, x];
             colInd = [colInd, y];
          end
       end

       newInput = input;

       for en = 1:numberOfBitError
          newInput(rowInd(en), colInd(en)) = newInput(rowInd(en),
          colInd(en)) * (-1);
       end
       testingData(ind).input = newInput;
       testingData(ind).name = name;
       ind = ind + 1;
   end
end
```

File #7

```
function trainingData = getHopfieldTrainingData

trainingData = [];

%%%%%%%%%%%%%%%%%%%%%%%%%%%%%%%
%% 2
%%%%%%%%%%%%%%%%%%%%%%%%%%%%%%%
trainingData(1).input = [...
    -1 -1 -1 -1 -1 -1 -1 -1;...
    -1 -1 -1 -1 -1 -1 -1 -1;...
     1  1  1  1  1 -1 -1  1;...
     1  1  1  1 -1 -1  1  1;...
     1  1  1 -1 -1  1  1  1;...
     1  1 -1 -1  1  1  1  1;...
     1 -1 -1 -1 -1 -1 -1 -1;...
     1 -1 -1 -1 -1 -1 -1 -1 ...
 ];
trainingData(1).name = '2';
%%%%%%%%%%%%%%%%%%%%%%%%%%%%%%%
%% 1
%%%%%%%%%%%%%%%%%%%%%%%%%%%%%%%
trainingData(2).input = [...
    1 1 1 -1 -1 1 1 1;...
    1 1 1 -1 -1 1 1 1;...
    1 1 1 -1 -1 1 1 1;...
    1 1 1 -1 -1 1 1 1;...
    1 1 1 -1 -1 1 1 1;...
    1 1 1 -1 -1 1 1 1;...
    1 1 1 -1 -1 1 1 1;...
    1 1 1 -1 -1 1 1 1 ...
 ];
trainingData(2).name = '1';
%%%%%%%%%%%%%%%%%%%%%%%%%%%%%%%
%% 4
%%%%%%%%%%%%%%%%%%%%%%%%%%%%%%%
trainingData(3).input = [...
    -1 -1  1  1  1  1  1 1;...
    -1 -1  1  1  1  1  1 1;...
    -1 -1  1  1 -1 -1  1 1;...
    -1 -1  1  1 -1 -1  1 1;...
    -1 -1 -1 -1 -1 -1 -1 1;...
    -1 -1 -1 -1 -1 -1 -1 1;...
     1  1  1  1 -1 -1  1 1;...
     1  1  1  1 -1 -1  1 1 ...
 ];
trainingData(3).name = '4';
```

```
%%%%%%%%%%%%%%%%%%%%%%%%%%%%%%%%
%% 0
%%%%%%%%%%%%%%%%%%%%%%%%%%%%%%%%
trainingData(4).input = [...
     1 -1 -1 -1 -1 -1 -1  1;...
     1 -1 -1 -1 -1 -1 -1  1;...
     1 -1 -1  1  1 -1 -1  1;...
     1 -1 -1  1  1 -1 -1  1;...
     1 -1 -1  1  1 -1 -1 -1;...
     1 -1 -1  1  1 -1 -1 -1;...
     1 -1 -1 -1 -1 -1 -1  1;...
     1 -1 -1 -1 -1 -1 -1  1 ...
  ];
trainingData(4).name = '0';
```

7.B. Hopfield Network Case Study†: Traveling Salesman Problem

7.B.1. *Introduction*

The traveling salesman problem (TSP) is a classical optimization problem. It is a NP-complete (Non-deterministic Polynomial) problem. There is no algorithm for this problem, which gives a perfect solution. Thus any algorithm for this problem is going to be impractical with certain examples.

There are various neural network algorithms that can be used to try to solve such constrain satisfaction problems. Most solution have used one of the following methods

- Hopfield Network
- Kohonen Self-organizing Map
- Genetic Algorithm
- Simulated Annealing

Hopfield explored an innovative method to solve this combinatorial optimization problem in 1986. Hopfield-Tank algorithm [Hopfied and Tank, 1985] used the energy function to to efficiently implement TSP. Many other NN algorithms then followed.

The TSP Problem: A salesman is required to visit each of a given set of cities once and only once, returning to the starting city at the end of his trip (or tour). The path that the salesman takes is called a tour. The tour of minimum distance is desired.

Assume that we are given n cities and a nonnegative integer distance D_{ij} between any two cities i and j. We are asked to find the shortest tour of the cities. We can solve this problem by enumerating all possible solutions, computing the cost of each and finding the best. Testing every possibility for an n city tour would require $n!$ (There are actually $(n-1)!/2$ calculations to consider) math additions. A 30 city

---

†Case study by Padmagandha Sahoo, ECE Dept., University of Illinois, Chicago, 2003.

tour would require 2.65×10^{32} additions. The amount of computation will increase dramatically with the increase in the number of cities.

The neural network approach tends to give solutions with less computing time than other available algorithms.

For n cities to be visited, let X_{ij} be the variable that has value 1 if the salesman goes from city i to city j and value 0, otherwise. Let D_{ij} be the distance from city i to city j. The TSP can also be stated as follows:

Minimize the linear objective function:

$$\sum_{i=j}^{n} \sum_{j=1}^{n} X_{ij} D_{ij}$$

A simple strategy for this problem is to numerate all feasible tours to calculate the total distance for each tour, and to pick the tour with the smallest total distance. However, if there are n cities in the tour, the number of all feasible tours would be $(n-1)!$. So this simple strategy becomes impractical if the number of cities is large. For example, if there are 11 cities to be visited, there will be $10! = 3{,}628{,}800$ possible tours (including the tour with the same route but the different direction). This number grows to over 6.2 billion with only 13 cities in the tour. Hence, Hopfield-Tank algorithm is used to approximately solve this problem with minimal computation. Few applications of TSP include determining a Postal Delivery network, find the optimal path for a school bus route etc.

7.B.2. *Hopfield neural network design*

The Hopfield network is a dynamic network, which iterates to converge from an arbitrary input state, as shown in Fig. 7.B.1. The Hopfield Network serves to minimize an energy function. It is a fully connected weighted network where the output of the network is fed back and there are weights to each of this link. A fully connected Hopfield network is shown in Fig. 7.B.1. Here we use n^2 neurons in the network, where n is the total number of cities to be visited. The neurons here have a threshold and step function. The inputs are given to the weighted input node. The major task is to find appropriate connection weights such that invalid tours should be prevented and valid tours should be preferred.

The output result of TSP can be represented in form of a Tour Matrix as in Fig. 7.B.2 below. The example is shown for 4 cities. The optimal visiting route, in the above example is:

$$\text{City2} \rightarrow \text{City1} \rightarrow \text{City4} \rightarrow \text{City3} \rightarrow \text{City2}$$

Hence, the total traveling distance is:

$$D = D_{21} + D_{14} + D_{43} + D_{32}$$

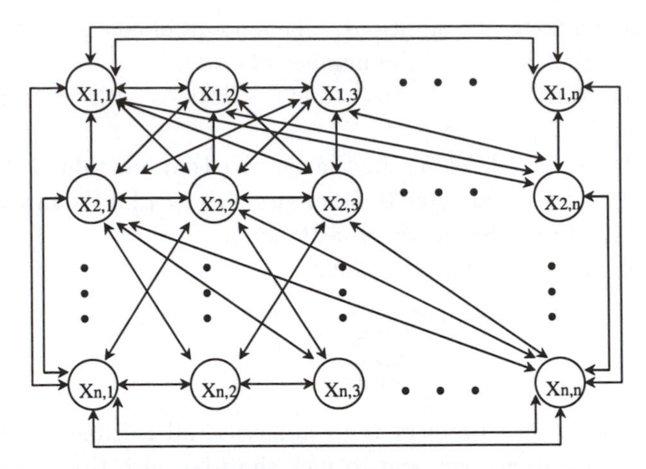

Fig. 7.B.1. The layout of Hopfield Network structure for TSP with n cities.

	#1	#2	#3	#4
C1	0	1	0	0
C2	1	0	0	0
C3	0	0	0	1
C4	0	0	1	0

Fig. 7.B.2. Tour matrix at network output.

7.B.2.1. *The energy function*

The Hopfield network for the application of the neural network can be best understood by the energy function. The energy function developed by Hopfield and Tank [1] is used for the project. The energy function has various hollows that represent the patterns stored in the network. An unknown input pattern represents a particular point in the energy landscape and the pattern iterates its way to a solution, the point moves through the landscape towards one of the hollows. The iteration is carried on for some fixed number of times or till the stable state is reached when the energy difference between two successive iterations lies below a very small threshold value (~ 0.000001).

The energy function used should satisfy the following criteria:

- The energy function should be able to lead to a stable combination matrix.
- The energy function should lead to the shortest traveling path.

The energy function used for the hopfield neural network is:

$$E = A\Sigma_i\Sigma_k\Sigma_{j\neq k}X_{ik}X_{ij} + B\Sigma_i\Sigma_k\Sigma_{j\neq k}X_{ki}X_{ji} + C[(\Sigma_i\Sigma_k X_{ik}) - n]_2$$
$$+ D\Sigma_k\Sigma_{j\neq k}\Sigma_i d_{kj}X_{ki}(X_{j,i+1} + X_{j,i-1})\,. \quad (1)$$

Here, A, B, C, D are positive integers. The setting of these constants are critical for the performance of Hopfield network. X_{ij} is the variable to denote the fact that city i is the jth city visited in a tour. Thus X_{ij} is the output of the jth neuron in the array of neurons corresponding to the ith city. We have n^2 such variables and their value will finally be 0 or 1 or very close to 0 or 1.

Hopfield and Tank [1] showed that if a combinatorial optimization problem can be expressed in terms of an energy function of the general form given in Eq. (1), a Hopfield network can be used to find locally optimal solutions of the energy function, which may translate to local minimum solutions of the optimization problem. Typically, the network energy function is made equivalent to the objective function which is to be minimized, while each of the constraints of the optimization problem are included in the energy function as penalty terms [4]. Sometimes a minimum of the energy function does not necessarily correspond to a constrained minimum of the objective function because there are likely to be several terms in the energy function, which contributes to many local minima. Thus, a tradeoff exists between which tems will be minimized completely, and feasibility of the network is unlikely unless the penalty parameters are chosen carefully. Furthermore, even if the network does manage to converge to a feasible solution, its quality is likely to be poor compared to other techniques, since the Hopfield network is a descent technique and converges to the first local minimum it encounters.

The energy function can be analyzed as follows:

- ***ROW CONSTRAINT***: $(A\Sigma_i\Sigma_k\Sigma_{j\neq k}X_{ik}X_{ij})$ In the energy function the first triple sum is zero if and only if there is only one "1" in each order column. Thus this takes care that no two or more cities are in same travel order i.e. no two cities are visited simultaneously.
- ***COLUMN CONSTRAINT***: $(B\Sigma_i\Sigma_k\Sigma_{j\neq k}X_{ki}X_{ji})$ In the energy function, the first triple sum is zero if and only if there is only one city appears in each order column. Thus this takes care that each city is visited only once.
- ***TOTAL NUMBER OF "1" CONSTRAINT***: $(C[(\Sigma_i\Sigma_k X_{ik}) - n]^2)$ The third triple sum is zero if and only if there are only N number of 1 appearing in the whole $n * n$ matrix. Thus this takes into care that all cities are visited.
- The first three summation are set up to satisfy the condition 1, which is necessary to produce a legal traveling path.
- ***SHORTEST DISTANCE CONSTRAINT***: $[D\Sigma_k\Sigma_{j\neq k}\Sigma_i d_{kj}X_{ki}(X_{j,i+1} + X_{j,i-1})]$ The forth triple summation provides the constrain for the shortest path. D_{ij} is the distance between city i and city j. The value of this term is minimum when the total distance traveled is shortest.
- The value of D is important to decide between the time taken to converge and the optimality of the solution. If the value of D is low it takes long time for the NN to converge but it gives solution nearer to the optimal solution but if the value of D is high, the network converges fast but the solution may not be optimal.

7.B.2.2. *Weight matrix setting*

The network here is fully connected with feedback and there are n^2 neurons, thus the weight matrix will be a square matrix of $n^2 * n^2$ elements. According to the Energy function the weight matrix can be set up as follows [1]:

$$W_{ik,lj} = -A\delta_{il}(1-\delta_{kj}) - B\delta_{kj}(1-\delta_{jl}) - C - Dd_{jl}(\delta_{j,k+1} + \delta_{j,k-1}) \tag{2}$$

Here the value of constants A, B, C, D are same as we have it in the Energy function. Weights are also updated keeping in mind various constraints to give a valid tour with minimum cost of travel. In this context, the Kronecker delta function (δ) is used to facilitate simple notation.

The weight function can be analyzed as follows:

- The neuron whose weight is updated is referred with two subscripts, one for the city it refers to and the other for the order of the city in the tour.
- Therefore, an element of the weight matrix for a connection between two neurons needs to have four subscripts, with a comma after two of the subscripts.
- The negative signs indicate inhibition through the lateral connections in a row or a column.
- The Kronecker delta function has two arguments (two subscripts of the symbol δ). By definition δ_{ik} has value 1 if $i = k$, and 0 if $i \neq k$.
- The first term gives the row constraint, thus taking care that no two cities are updated simultaneously.
- The second term gives the column constraint, thus taking care that no city is visited more than once.
- The third term here is for global inhibitation
- The fourth term takes care of the minimum distance covered.

7.B.2.3. *Activation function*

The activation function also follows various constraints to get a valid path. It can be defined as follows [1]:

$$\begin{aligned}
a_{ij} &= \Delta t(T_1 + T_2 + T_3 + T_4 + T_5) \\
T_1 &= -a_{ij}/\tau \\
T_2 &= -A\Sigma_i X_{ik} \\
T_3 &= -B\Sigma_i X_{ik} \\
T_4 &= -C(\Sigma_i\Sigma_k\Sigma_{ik} - m) \\
T_5 &= -D\Sigma_k d_{ik}(X_{k,j+1} + X_{k,j-1})
\end{aligned} \tag{3}$$

- We denote the activation of the neuron in the ith row and jth column by a_{ij}, and the output is denoted by X_{ij}.
- A time constant τ is also used. The value of τ is taken as 1.0.
- A constant m is also another parameter used. The value of **m** is 15.

- The first term in activation function is decreasing on each iteration.
- The second, third, fourth and the fifth term give the constraints for the valid tour.

The activation is updated as:

$$a_{ij}(\text{new}) = a_{ij}(\text{old}) + \Delta a_{ij} \,. \tag{4}$$

7.B.2.4. *The activation function*

This a continuous hopfield network with the following output function

$$X_{ij} = (1 + \tanh(\lambda a_{ij}))/2 \,. \tag{5}$$

- Here X_{ij} is the output of the neuron.
- The hyperbolic tangent function gives an output.
- The value of λ determines the slope of the function. Here the value of λ is 3.
- Ideally we want output either 1 or 0. But the hyperbolic tangent function gives a real number and we settle at a value that is very close to desired result, for example, 0.956 instead of 1 or say 0.0078 instead of 0.

7.B.3. *Input selection*

The inputs to the network are chosen arbitrarily. The initial state of the network is thus not fixed and is not biased against any particular route. If as a consequence of the choice of the inputs, the activation works out to give outputs that add up to the number of cities, and initial solution for the problem, a legal tour will result. A problem may also arise that the network will get stuck to a local minimum. To avoid such an occurrence, random noise is generated and added to initial input.

Also there are inputs that are taken from user. The user is asked to input the number of cities he want to travel and the distance between those cities which are used to generate the distance matrix.

Distance matrix is a n^2 square matrix whose principal diagonal is zero. Figure 7.B.3 below shows a typical distance matrix for 4 cities.

	C1	C2	C3	C4
C1	0	10	18	15
C2	10	0	13	26
C3	18	13	0	23
C4	15	26	23	0

Fig. 7.B.3. Distance matrix (based on distance information input).

Hence, the distance between cities C1 and C3 is 18 and distance of a city to itself is 0.

7.B.4. *Implementation details*

The algorithm is implemented in C++ for the Hopfield network operation for the traveling salesman problem. This code can handle for maximum upto 25 cities it can be very easily extended for more number of cities. The following steps are followed to implement this network.

1. Given the number of N cities and their co-ordinates, compute the Distance matrix D.
2. Initialize the network and setup the weight matrix as shown in Eq. (2).
3. Randomly assign initial input states to the network and compute the activation and output of the network, After that, the network is left alone, and it proceeds to cycle through a succession of states, until it converges to a stable solution.
4. Compute the energy using Eq. (1) for each iteration. Energy should decrease from iteration to iteration.
5. Iterate the updating to the activation and output until the network converges to a stable solution. This happens when the change in energy between two successive iterations lies below a small threshold value (~ 0.000001) or, when the energy starts to increase instead of decreasing.

The following is a listing of the characteristics of the C++ program along with definitions and/or functions used. The number of cities and the distances between the cities are solicited from the user.

- The distance is taken as integer values.
- A neuron corresponds to each combination of a city and its order in the tour. The ith city visited in the order j, is the neuron corresponding to the element $j + i^*n$, in the array for neurons. Here n is the number of cities. The i and j vary from 0 to n^{-1}. There are n^2 neurons.
- A random input vector is generated in the function main( ), and is later referred to as the input activation for the network.
- getnwk( ): It generates the weight matrix as per Eq. (2). It is a square matrix of order n^2.
- initdist( ): It takes the distances between corresponding cities from the user given in form of distance matrix.
- asgninpt( ): It assigns the randomly generated intial input to the network.
- iterate( ): This function finds the final path that is optimal or near optimal. It iterates and the final state of the network is set in such a way that all the constraint of the network is fulfilled.
- Getacts( ): Compute the output of the activation function that gets used in Iterate( ) routine.
- findtour( ): It generates a Tour Matrix and the exact route of travel.
- calcdist( ): calculates the total distance of the tour based on the tour generated by the function findtour( ).

Parameters setting

The parameter settings in the Hopfield network are critical to the performance of the Network. The initial values of the input parameters used are as follows:

$$A : 0.5$$
$$B : 0.5$$
$$C : 0.2$$
$$D : 0.5$$
$$\lambda : 3.0$$
$$\tau : 1.0$$
$$m : 15$$

7.B.5. *Output results*

The attached result shows the simulation using 5, 10, 15, 20, 25 cities. The traveling paths generated are shown in form of matrices, which are in the "output.txt" file.

Hopfield neural network is efficient and can converge to stable states in a finite number of iterations. It is observed that for upto 20 cities problem, the network converges well to a stable state most of the time with a global minimum solution. However, with further increase in the number of cities, the network converges to a stable state less frequently. The graphical outputs are shown in *Appendix 1*.

The ***output.txt*** file (*Appendix 2*) first gives the inputs that are taken from the user i.e. the number of cities and their distances in the form of distance matrix. Then for those cities the output that is generated is printed in the form of Tour Matrix, Tour Route and Total Distance Traveled. The solution is optimal or near optimal.

The results attached along with the code are for 5, 10, 15, 20, 25 cities respectively. The number of iterations and time taken to converge the network in each case can be summarized as follows:

Cities	Iteration	Time (sec)	Result
5	152	0.4652	Good
10	581	1.8075	Good
15	1021	3.2873	Good
20	2433	7.6458	Good
25	5292	16.2264	OK

The **graphical output** representations of routes for 5, 10, 15, 20 and 25 citlies are shown in Fig. 7.B.4 below. Figure 7.B.5 illustrates the energy convergenge for the 5, 10, 15, 20 cities problems and Fig. 7.B.6 shows the number of iterations required for convergence vs. number of cities.

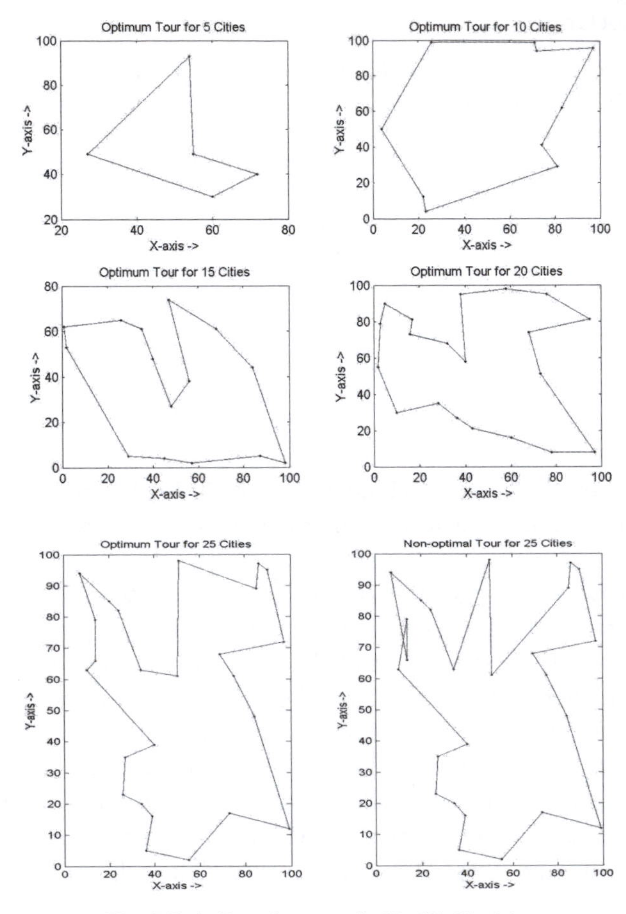

Fig. 7.B.4. Travel route: 5, 10, 15, 20 cities.

Comments on the results:

- The result above shows that as the number of cities increases the number of iteration required increases sharply. The increase is not a linear increase.
- The number of iterations required for the convergence did not remain same for any particular city. For example, for 5 cities the network usually converged between 120 to 170 iterations, but on few occasions it took around 80 iterations while in few cases it did not converge at all or took more than 250 iterations. This is because the initial network state is randomly generated. This may sometimes result to no convergence also.
- Many times the result converges to local minimum instead of global minimum. To avoid this, random bias is added to the initial inputs.
- The algorithm developed is non-deterministic. Thus it does not promise an optimal solution every time. Though it does give near optimal solution in most of the cases, it may fail to converge and give a correct solution.

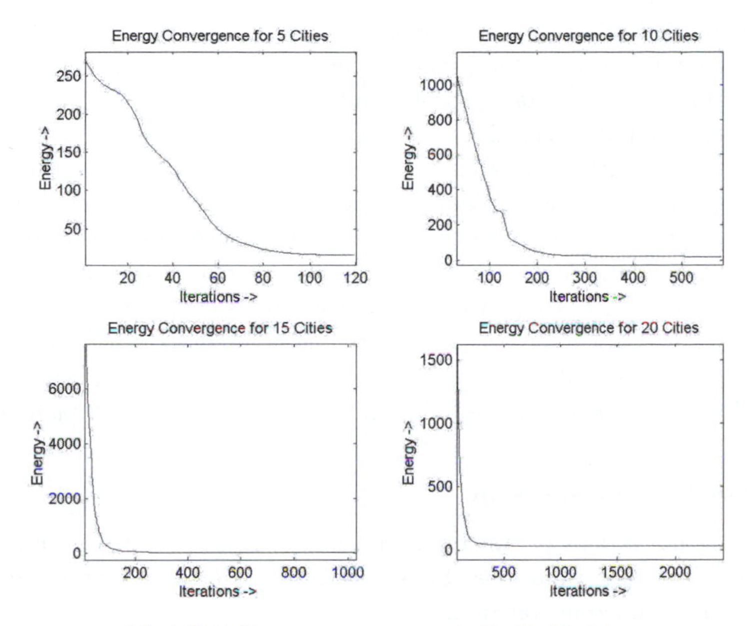

Fig. 7.B.5. Energy convergence: 5, 10, 15, 20 cities.

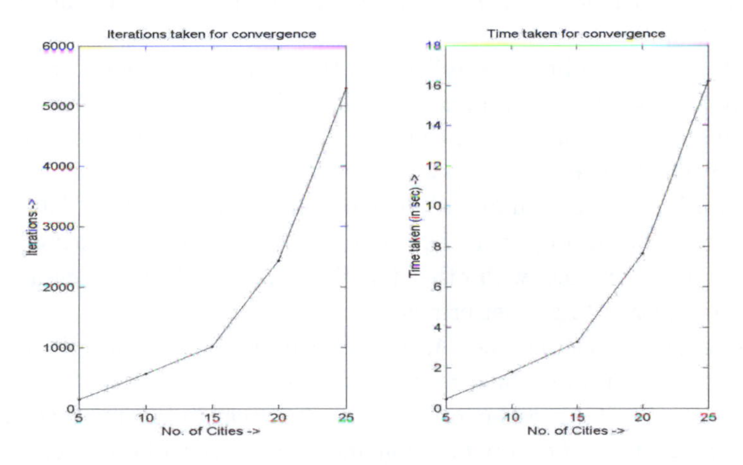

Fig. 7.B.6. Number of iterations required for convergence vs. number of cities.

- Many times when the energy of the system was calculated, it was found to increase instead of decreasing.
- Thus the algorithm failed in few cases. This again was the consequence of the random initial state of the network.
- In 93% of test cases the algorithm converged, while in 7% algorithm failed to converge and sometimes the energy of the system increased instead of decreasing while the network iterates towards convergence. There are various advantages of using Hopfield network though I had seen many other approaches like Kohonen

Network and Genetic Algorithm approach.

- Hopfield neural network setup is very optimal for the solution of TSP. It can be easily used for the optimization problems like that of TSP.
- It gives very accurate result due to very powerful and complete Energy equation developed by Hopfield and Tank.
- The approach is much faster than Kohonen as the number of iteration required to get the solution is less. The result obtained is much more near optimal than compared to Genetic Algorithm approach as in genetic algorithm it is more like trial error and chances to get the optimal solution is comparatively very less.
- This neural network approach is very fast compared to standard programing techniques used for TSP solution. With very few changes this algorithm can be modified to get the approximate solution for many other NP-complete problems.

7.B.6. *Concluding discussion*

- The setting for various parameter values like $A, B, C, D, \lambda, \tau, m$, etc is a major challenge. The best value was chosen by trial and error. Improvement is still possible for these parameter values.
- Many times the algorithm converged to local minima instead of global minimum. This problem was mostly resolved by adding a random noise to the initial inputs of the system [2].
- The testing of algorithm gets difficult as the number of cities increase. Though there are few software and programs available for the testing, none of them guarantees the optimal solution each time. So an approximation was made during the testing of the algorithm.
- The network, as developed below, does not always give optimal solution though in most cases it is near optimal. Few more changes or improvements can be made to energy function along with other functions like weight updating function and activation function to get better answer.
- Various values of constants (i.e. A, B, C, D) can be tried in multiple combinations to get optimal or near optimal result in the present algorithm.
- Problems of infeasibility and poor solution quality can be essentially eliminated by an appropriate form of energy function and modification of the internal dynamics of the Hopfield network. By expressing all constraints of the problem in a single term, overall number of terms and parameters in the energy function can be reduced [4].
- Even if one of the distances between the cities is wrong the network has to start form the very first stage. This error can be handled in some way in future.
- If we want to add or delete a city, the network must be restarted from the initial state with the required changes. Some equations can be developed to incorporate these changes.
- The algorithm can be modified for solving other NP-complete problems.

7.B.7. *Source code (C++)*

```
//TSP.CPP
#include "tsp.h"
#include <stdlib.h>
#include <time.h>

int randomnum(int maxval) // Create random numbers between 1 to 100
{
         return rand()%maxval;
}

/* ========= Compute the Kronecker delta function ====================== */
int krondelt(int i,int j)
{
        int k;
        k=((i==j)?(1):(0));
        return k;
};

/* ========== compute the distance between two co-ordinates =============== */
int distance(int x1,int x2,int y1,int y2)
{
        int x,y,d;

        x=x1-x2;
        x=x*x;
        y=y1-y2;
        y=y*y;
        d=(int)sqrt(x+y);

        return d;
}


void neuron::getnrn(int i,int j)
{
        cit=i;
        ord=j;
        output=0.0;
        activation=0.0;
}

/* =========== Randomly generate the co-ordinates of the cities ================== */

void HP_network::initdist(int cityno) //initiate the distances between the k cities
{
        int i,j;
        int rows=cityno, cols=2;
        int **ordinate;
        int **row;

        ordinate = (int **)malloc((rows+1) *sizeof(int *));/*one extra for sentinel*/

/* now allocate the actual rows */
```

```
for(i = 0; i < rows; i++)
        {
  ordinate[i] = (int *)malloc(cols * sizeof(int));
}

/* initialize the sentinel value */
ordinate[rows] = 0;

srand(cityno);

      for(i=0; i<rows; i++)
      {
               ordinate[i][0] = rand() % 100;
               ordinate[i][1] = rand() % 100;
      }

      outFile<<"\nThe Co-ordinates of "<<cityno<<" cities: \n";

      for (i=0;i<cityno;++i)
      {
               outFile<<"X "<<i<<": "<<ordinate[i][0]<<"  ";
               outFile<<"Y "<<i<<": "<<ordinate[i][1]<<"\n";
      }

for (i=0;i<cityno;++i)
      {
               dist[i][i]=0;
               for (j=i+1;j<cityno;++j)
               {
                dist[i][j]=distance(ordinate[i][0],ordinate[j][0],
                ordinate[i][1],ordinate[j][1])/1;
               }
      }

      for (i=0;i<cityno;++i)
      {
               for (j=0;j<i;++j)
               {
                        dist[i][j]=dist[j][i];
               }
      }

      print_dist();     //print the distance matrix
      cout<<"\n";

      for(row = ordinate; *row != 0; row++)
      {
   free(*row);
}

      free(ordinate);
}

/* ============== Print Distance Matrix =================== */
```

```
void HP_network::print_dist()
{
        int i,j;
        outFile<<"\n Distance Matrix\n";

        for (i=0;i<cityno;++i)
        {
                for (j=0;j<cityno;++j)
                {
                        outFile<<dist[i][j]<<"  ";

                }
                outFile<<"\n";
         }
}

/* ============ Compute the weight matrix ==================== */

void HP_network::getnwk(int citynum,float x,float y,float z,float w)
{
         int i,j,k,l,t1,t2,t3,t4,t5,t6;
         int p,q;
         cityno=citynum;
         a=x;
         b=y;
         c=z;
         d=w;
         initdist(cityno);

         for (i=0;i<cityno;++i)
         {
                 for (j=0;j<cityno;++j)
                 {
                         tnrn[i][j].getnrn(i,j);
                 }
         }

         for (i=0;i<cityno;++i)
         {
                 for (j=0;j<cityno;++j)
                 {
                         p=((j==cityno-1)?(0):(j+1));
                         q=((j==0)?(cityno-1):(j-1));
                         t1=j+i*cityno;
                         for (k=0;k<cityno;++k)
                         {
                                 for (l=0;l<cityno;++l)
                                 {
                                         t2=l+k*cityno;
                                         t3=krondelt(i,k);
                                         t4=krondelt(j,l);
                                         t5=krondelt(l,p);
                                         t6=krondelt(l,q);
                                         weight[t1][t2]=-a*t3*(1-t4)-b*t4*(1-t3)
-c-d*dist[i][k]*(t5+t6)/100;
```

```
                                        }
                                }
                        }
                }

//              print_weight(cityno);
}

void HP_network::print_weight(int k)
{
                int i,j,nbrsq;
                nbrsq=k*k;
                cout<<"\nWeight Matrix\n";
                outFile<<"\nWeight Matrix\n";
                for (i=0;i<nbrsq;++i)
                {
                        for (j=0;j<nbrsq;++j)
                        {
                                outFile<<weight[i][j]<<" ";
                        }
                        outFile<<"\n";
                }
}

/* =========== Assign initial inputs to the network ============= */

void HP_network::asgninpt(float *ip)
{
                int i,j,k,l,t1,t2;

                for (i=0;i<cityno;++i)
                {
                        for (j=0;j<cityno;++j)
                        {
                                acts[i][j]=0.0;
                        }
                }

                //find initial activations
                for (i=0;i<cityno;++i)
                {
                        for (j=0;j<cityno;++j)
                        {
                                t1=j+i*cityno;
                                for (k=0;k<cityno;++k)
                                {
                                        for (l=0;l<cityno;++l)
                                        {
                                                t2=l+k*cityno;
                                                acts[i][j]+=weight[t1][t2]*ip[t1];
                                        }
                                }
                        }
                }
```

```
            //print activations
//          outFile<<"\n initial activations\n";
//          print_acts();
}

/* ======== Compute the activation function outputs =================== */

void HP_network::getacts(int nprm,float dlt,float tau)
{
            int i,j,k,p,q;
            float r1,r2,r3,r4,r5;
            r3=totout-nprm;

            for (i=0;i<cityno;++i)
            {
                        r4=0.0;
                        p=((i==cityno-1)?(0):(i+1));
                        q=((i==0)?(cityno-1):(i-1));
                        for (j=0;j<cityno;++j)
                        {
                                    r1=citouts[i]-outs[i][j];
                                    r2=ordouts[j]-outs[i][j];
                                    for (k=0;k<cityno;++k)
                                    {
                                                r4+=dist[i][k]*(outs[k][p]+outs[k][q])/100;
                                    }
                                    r5=dlt*(-acts[i][j]/tau-a*r1-b*r2-c*r3-d*r4);
                                    acts[i][j]+=r5;
                         }
            }

}

/* ============== Get Neural Network Output ===================== */

void HP_network::getouts(float la)
{
            double b1,b2,b3,b4;
            int i,j;
            totout=0.0;

            for (i=0;i<cityno;++i)
            {
                        citouts[i]=0.0;
                        for (j=0;j<cityno;++j)
                        {
                                    b1=la*acts[i][j];
                                    b4=b1;
                                    b2=exp(b4);
                                    b3=exp(-b4);
                                    outs[i][j]= (float)(1.0+(b2-b3)/(b2+b3))/2.0;
                                    citouts[i]+=outs[i][j];
                         }
                         totout+=citouts[i];
             }
```

```
            for (j=0;j<cityno;++j)
            {
                    ordouts[j]=0.0;
                    for (i=0;i<cityno;++i)
                    {
                            ordouts[j]+=outs[i][j];
                    }
            }
}

/* ============ Compute the Energy function ====================== */

float HP_network::getenergy()
{
   int i,j,k,p,q;
           float t1,t2,t3,t4,e;

           t1=0.0;
           t2=0.0;
           t3=0.0;
           t4=0.0;
           for (i=0;i<cityno;++i)
           {
                   p=((i==cityno-1)?(0):(i+1));
                   q=((i==0)?(cityno-1):(i-1));

                   for (j=0;j<cityno;++j)
                   {
                           t3+=outs[i][j];
                           for (k=0;k<cityno;++k)
                           {
                                   if (k!=j)
                                   {
                                           t1+=outs[i][j]*outs[i][k];
                                           t2+=outs[j][i]*outs[k][i];
                                           t4+=dist[k][j]*outs[k][i]
                                           *(outs[j][p]+outs[j][q])/10;
                                   }
                            }
                   }
           }

     t3=t3-cityno;
           t3=t3*t3;
           e=0.5*(a*t1+b*t2+c*t3+d*t4);

           return e;
}

/* ======== find a valid tour ======================== */

void HP_network::findtour()
{
       int i,j,k,tag[Maxsize][Maxsize];
```

```
        float tmp;
        for (i=0;i<cityno;++i)
        {
                for (j=0;j<cityno;++j)
                {
                        tag[i][j]=0;
                }
        }

        for (i=0;i<cityno;++i)
        {
                tmp=-10.0;
                for (j=0;j<cityno;++j)
                {
                        for (k=0;k<cityno;++k)
                        {
                                if ((outs[i][k]>=tmp)&&(tag[i][k]==0))
                                        tmp=outs[i][k];
                        }

                        if ((outs[i][j]==tmp)&&(tag[i][j]==0))
                        {
                                tourcity[i]=j;
                                tourorder[j]=i;
        cout<<"tour order"<<j<<"\n";
                                for (k=0;k<cityno;++k)
                                {
                                        tag[i][k]=1;
                                        tag[k][j]=1;
                                }
                        }
                }
        }
}

//print outputs
void HP_network::print_outs()
{
        int i,j;
        outFile<<"\n the outputs\n";
        for (i=0;i<cityno;++i)
        {
                for (j=0;j<cityno;++j)
                {
                        outFile<<outs[i][j]<<" ";
                }
                outFile<<"\n";
        }
}

/* ======= Calculate total distance for tour ============== */

void HP_network::calcdist()
{
        int i,k,l;
```

```
        distnce=0.0;

        for (i=0;i<cityno;++i)
        {
                k=tourorder[i];
                l=((i==cityno-1)?(tourorder[0]):(tourorder[i+1]));
                distnce+=dist[k][l];
        }

        outFile<<"\nTotal distance of tour is : "<<distnce<<"\n";
}

/* ======= Print Tour Matrix ============================== */

void HP_network::print_tour()
{
        int i;
        outFile<<"\nThe tour order: \n";

        for (i=0;i<cityno;++i)
        {
                outFile<<tourorder[i]<<" ";
                outFile<<"\n";
        }
}

/* ======= Print network activations ======================= */

void HP_network::print_acts()
{
        int i,j;
        outFile<<"\n the activations:\n";
        for (i=0;i<cityno;++i)
        {
                for (j=0;j<cityno;++j)
                {
                        outFile<<acts[i][j]<<" ";
                }
                outFile<<"\n";
        }
}

/*========== Iterate the network specified number of times =============== */

void HP_network::iterate(int nit,int nprm,float dlt,float tau,float la)
{
        int k,b;
  double oldenergy,newenergy, energy_diff;
        b=1;
        oldenergy=getenergy();
        outFile1<<""<<oldenergy<<"\n";
        k=0;
        do
        {
                getacts(nprm,dlt,tau);
```

```
                getouts(la);
                newenergy=getenergy();
                outFile1<<""<<newenergy<<"\n";

                //energy_diff = oldenergy - newenergy;
                //if (energy_diff < 0)
                //          energy_diff = energy_diff*(-1);

                if (oldenergy - newenergy < 0.0000001)
                {
                         //printf("\nbefore break: %lf\n", oldenergy - newenergy);
                         break;
                }

                oldenergy = newenergy;
                k++;
        }
        while (k<nit) ;

        outFile<<"\n"<<k<<" iterations taken for convergence\n";
        //print_acts();
        //outFile<<"\n";
        //print_outs();
        //outFile<<"\n";

}


void main()
{

/*========= Constants used in Energy, Weight and Activation Matrix ============== */

        int nprm=15;
  float a=0.5;
        float b=0.5;
        float c=0.2;
        float d=0.5;
        double dt=0.01;
        float tau=1;
        float lambda=3.0;
        int i,n2;
        int numit=4000;
        int cityno=15;
//      cin>>cityno;  //No. of cities
        float input_vector[Maxsize*Maxsize];
  time_t start,end;
        double dif;

        start = time(NULL);
        srand((unsigned)time(NULL));
        //time (&start);

  n2=cityno*cityno;
        outFile<<"Input vector:\n";
```

```
         for (i=0;i<n2;++i)
         {
                  if (i%cityno==0)
                  {
                      outFile<<"\n";
                  }
                  input_vector[i]=(float)(randomnum(100)/100.0)-1;
                  outFile<<input_vector[i]<<"  ";
          }

          outFile<<"\n";

//creat HP_network and operate

        HP_network *TSP_NW=new HP_network;
        if (TSP_NW==0)
        {
                cout<<"not enough memory\n";
                exit(1);
        }

        TSP_NW->getnwk(cityno,a,b,c,d);
        TSP_NW->asgninpt(input_vector);
        TSP_NW->getouts(lambda);
        //TSP_NW->print_outs();
  TSP_NW->iterate(numit,nprm,dt,tau,lambda);
        TSP_NW->findtour();
        TSP_NW->print_tour();
        TSP_NW->calcdist();
        //time (&end);
        end = time(NULL);
  dif = end - start;
        printf("Time taken to run this simulation: %lf\n",dif);

}

/******************************************************************************

        Network:     Solving TSP using Hopfield Network
                           ECE 559  (Neural Networks)

        Author:       PADMAGANDHA SAHOO
        Date:          11th Dec '03

 ******************************************************************************/
// TSP.H

#include <iostream.h>
#include <stdlib.h>
#include <math.h>
#include <stdio.h>
#include <time.h>
#include <fstream.h>

#define Maxsize 30
```

```
ofstream outFile("Output.txt",ios::out);
ofstream outFile1("Output1.txt",ios::out);

class neuron
{
protected:
        int cit,ord;
        float output;
        float activation;
        friend class HP_network;

public:
        neuron() {};
        void getnrn(int,int);
};

class HP_network
{
public:
        int cityno;         //Number of City
        float a,b,c,d,totout,distnce;

        neuron (tnrn)[Maxsize][Maxsize];
        int dist[Maxsize][Maxsize];
        int tourcity[Maxsize];
        int tourorder[Maxsize];
        float outs[Maxsize][Maxsize];
        float acts[Maxsize][Maxsize];
        float weight[Maxsize*Maxsize][Maxsize*Maxsize];
        float citouts[Maxsize];
        float ordouts[Maxsize];
        float energy;

        HP_network() { };
        void getnwk(int,float,float,float,float);
        void initdist(int);
        void findtour();
        void asgninpt(float *);
        void calcdist();
        void iterate(int,int,float,float,float);
        void getacts(int,float,float);
        void getouts(float);
        float getenergy();

        void print_dist();          // print the distance matrix among n cities
        void print_weight(int);     // print the weight matrix of the network
        void print_tour();          // print the tour order of n cities
        void print_acts();          // print the activations of the neurons
                                       in the network
        void print_outs();          // print the outputs of the neurons in
                                       the network
};
```

```
%%%%%%%%%%%%%%%%%%%%%%%%%%%%%%%%%%%%%%%%%%%%%%%%%%
% Matlab Routine: tour.m
% Description: This routine contains the code to plot all graphical outputs for the
%              TSP problem. It plots the optimum tour for all cities, energy con-
%              vergence graph, iterations and time taken for each simulation etc.
%%%%%%%%%%%%%%%%%%%%%%%%%%%%%%%%%%%%%%%%%%%%%%%%%%

clear all; close all;

x = [54 55 72 60 27 54];
y = [93 49 40 30 49 93];
subplot(2,2,1);plot(x,y,'.-');
title('Optimum Tour for 5 Cities');
xlabel('X-axis →');ylabel('Y-axis →');

x = [4 22 23 81 74 83 97 72 71 26 4];
y = [50 12 4 29 41 62 96 94 99 99 50];
subplot(2,2,2);plot(x,y,'.-');
title('Optimum Tour for 10 Cities');
xlabel('X-axis →');ylabel('Y-axis →');

x = [2 1 26 35 40 48 56 47 68 84 98 87 57 45 29 2];
y = [53 62 65 61 48 27 38 74 61 44 2 5 2 4 5 53];
subplot(2,2,3);plot(x,y,'.-');
title('Optimum Tour for 15 Cities');
xlabel('X-axis →');ylabel('Y-axis →');

x = [10 2 3 5 17 16 32 40 38 58 76 95 68 73 97 78 60 43 36 28 10];
y = [30 55 79 90 81 73 68 58 95 98 95 81 74 51 8 8 16 21 27 35 30];
subplot(2,2,4);plot(x,y,'.-');
title('Optimum Tour for 20 Cities');
xlabel('X-axis →');ylabel('Y-axis →');

x = [10 14 14 7 20 24 34 50 51 85 86 90 97 69 75 84 99 73 55 36 39 34 26 27 40 10];
y = [63 66 79 94 85 82 63 61 98 89 97 95 72 68 61 48 12 17 2 5 16 20 23 35 39 63];
figure;subplot(1,2,1);plot(x,y,'.-');
title('Optimum Tour for 25 Cities');
xlabel('X-axis →');ylabel('Y-axis →');

x = [10 14 14 7 20 24 34 50 51 85 86 90 97 69 75 84 99 73 55 36 39 34 26 27 40 10];
y = [63 79 66 94 85 82 63 98 61 89 97 95 72 68 61 48 12 17 2 5 16 20 23 35 39 63];
subplot(1,2,2);plot(x,y,'.-');
title('Non-optimal Tour for 25 Cities');
xlabel('X-axis →');ylabel('Y-axis →');

% Plot the graphs to show iterations and time taken for each simulation
iteration = [152 581 1021 2433 5292];
city = [5 10 15 20 25];
time = [0.4652 1.8075 3.2873 7.6458 16.2264];
figure;subplot(1,2,1);plot(city,iteration,'.-');
```

```
title('Iterations taken for convergence');
ylabel('Iterations →');xlabel('No. of Cities →');
subplot(1,2,2);plot(city,time,'.-');
title('Time taken for convergence');
ylabel('Time taken (in sec) →');xlabel('No. of Cities →');

% Plot the Energy convergence plots
n5 = textread('energy5.txt');
n10 = textread('energy10.txt');
n15 = textread('energy15.txt');
n20 = textread('energy20.txt');
n25 = textread('energy25.txt');
figure;subplot(2,2,1);plot(n5);
title('Energy Convergence for 5 Cities');
ylabel('Energy →');xlabel('Iterations →');
subplot(2,2,2);plot(n10);
title('Energy Convergence for 10 Cities');
ylabel('Energy →');xlabel('Iterations →');
subplot(2,2,3);plot(n15);
title('Energy Convergence for 15 Cities');
ylabel('Energy →');xlabel('Iterations →');
subplot(2,2,4);plot(n20);
title('Energy Convergence for 20 Cities');
ylabel('Energy →');xlabel('Iterations →');
```

Chapter 8

Counter Propagation

8.1. Introduction

The *Counter Propagation* neural network, due to Hecht–Nielsen (1987), is faster by approximately a factor of 100 than back propagation, but more limited in range of applications. It combines the Self-Organizing (*Instar*) networks of Kohonen (1984) and the Grossberg's *Oustar* net (1969, 1974, 1982) consisting of one layer of each. It has good properties of *Generalization* (essential, in some degree, to all neural networks) that allow it to deal well with *partially incomplete* or *partially incorrect input vectors.* Counter Propagation network serves as a very fast clustering network.

Its *Structure* is as in Fig. 8.1, where a (hidden) K-layer is followed by an output G-layer.

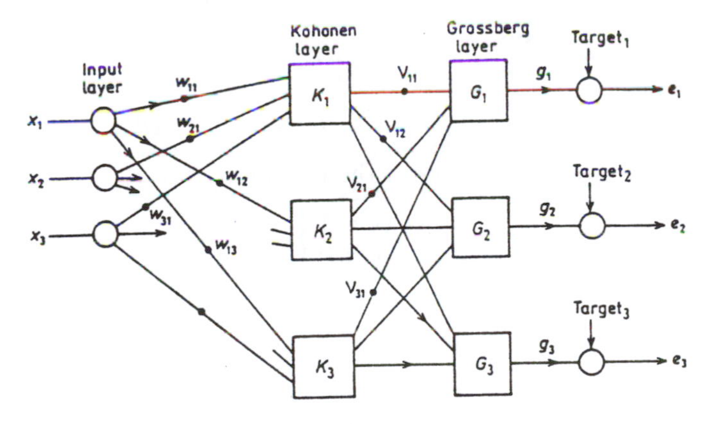

Note: number of neurons in the Kohonen and the Grossberg layer need not be equal.

Fig. 8.1. Counter propagation network.

8.2. Kohonen Self-Organizing Map (SOM) Layer

The Kohonen layer [Kohonen 1984, 1988] is a "Winner-take-all" (WTA) layer. Thus, for a *given input vector, only one Kohonen layer output is 1 whereas all others are 0.* No training vector is required to achieve this performance. Hence, the name: Self-Organizing Map Layer (SOM-Layer).

Let the net output of a Kohonen layer neuron be denoted as k_j. Then

$$k_j = \sum_{i=1}^{m} w_{ij} x_i = \mathbf{w}_i^T \mathbf{x}\,; \quad \mathbf{w}_j \triangleq [w_{1j} \cdots w_{mj}]^T$$
$$\mathbf{x} \triangleq [x_1 \cdots x_m]^T \tag{8.1}$$

and, for the hth ($j = h$) neuron where

$$k_h > k_{j \neq h} \tag{8.2}$$

we then set w_j such that:

$$k_h = \sum_{i=1}^{m} w_{ih} x_i = 1 = \mathbf{w}_h^T \mathbf{x} \tag{8.3a}$$

and, possibly via lateral inhibition as in Sec. 9.2.b

$$k_{j \neq h} = 0 \tag{8.3b}$$

8.3. Grossberg Layer

The output of the Grossberg layer is the weighted output of the Kohonen layers, by Fig. 8.1.

Denoting the net output of the Grossberg layer [Grossberg, 1974] as g_j then

$$g_j = \sum_i k_i v_{ij} = \mathbf{k}^T \mathbf{v}_j\,; \ \mathbf{k} \triangleq [k_1 \cdots k_m]^T$$
$$\mathbf{v}_j \triangleq [v_{1j} \cdots v_{mj}]^T \tag{8.4}$$

But, by the "winner-take-all" nature of the Kohonen layer; if

$$\left.\begin{array}{r} k_h = 1 \\ h_{j \neq h} = 0 \end{array}\right\} \tag{8.5}$$

then

$$g_j = \sum_{i=1}^{p} k_i v_{ij} = k_n v_{hj} = v_{hj} \tag{8.6}$$

the right-hand side equality being due to $k_h = 1$.

8.4. Training of the Kohonen Layer

The Kohonen layer acts as a *classifier* where all *similar input vectors*, namely those belonging to the same class produce a unity output in the same Kohonen neuron. Subsequently, the *Grossberg layer produces the desired output* for the given class as has been classified in the Kohonen layer above. In this manner, generalization is then accomplished.

8.4.1. *Preprocessing of Kohonen layer's inputs*

It is usually required to normalize the Kohonen layer's inputs, as follows

$$x'_i = \frac{x_i}{\sqrt{\sum_j x_j^2}} \tag{8.7}$$

yield a normalized input vector $\mathbf{x}'$ where

$$(\mathbf{x}')^T\mathbf{x}' = 1 = \|\mathbf{x}'\| \tag{8.8}$$

The training of the Kohonen layer now proceeds as follows:

1. Normalize the input vector $\mathbf{x}$ to obtain $\mathbf{x}'$
2. The Kohonen layer neuron whose

$$(\mathbf{x}')^T\mathbf{w}_h = k'_h5 \tag{8.9}$$

is the highest, is declared the winner and its weights are adjusted to yield a unity output $k_h = 1$

Note that

$$k'_h = \sum_i x'_i w_{ih} = x'_1 w_{h1} + x'_2 w_{h2} + \cdots x'_m w_{hm} = (\mathbf{x}')^T\mathbf{w}_h \tag{8.10}$$

But since

$$(\mathbf{x}')^T\mathbf{x}' = 1$$

and by comparing Eqs. (8.9) and (8.10) we obtain that

$$\mathbf{w} = \mathbf{x}' \tag{8.11}$$

namely, the weight vector of the winning Kohonen neuron (the hth neuron in the Kohonen layer) equals (best approximates) the input vector. *Note that* there is *no "teacher"*. We *start with* the winning weights *to be the ones that best approximate* $\mathbf{x}$ and then we make these weights *even more similar to* $\mathbf{x}$, via

$$\mathbf{w}(n+1) = \mathbf{w}(n) + \alpha\,[\mathbf{x} - \mathbf{w}(n)] \tag{8.12}$$

where α is a training rate coefficient (usually $\alpha \cong 0.7$) and it may be *gradually reduced* to allow large initial steps and smaller for final convergence to $\mathbf{x}$.

In case of a *single input* training vector, one can simply set the weight to *equal* the inputs *in a single step.*

If many training input-vectors of the same class are employed, *all of which are supposed to activate* the *same* Kohonen neuron, the weights should become the average of the input vectors $\mathbf{x}_i$ of a given class h, as in Fig. 8.2.

Since $\|w_{n+1}\|$ above is not necessarily 1, it must be normalized to 1 once derived as above!

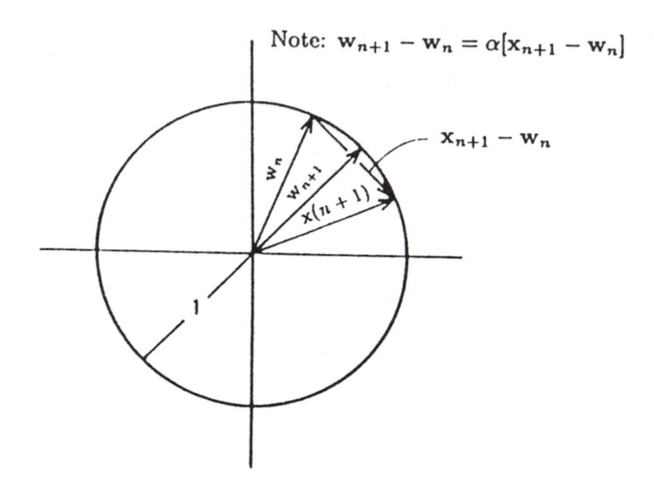

Fig. 8.2. Training of a Kohonen layer.

8.4.2. *Initializing the weights of the Kohonen layer*

Whereas in practically all NN's the initial weights are selected to be of *pseudo random low values*, in the case of Kohonen networks, any *pseudo random weights* must be *normalized* if an approximation to $\mathbf{x}'$ is to be of any meaning. *But then,* even normalized random weights may be too far off from $\mathbf{x}'$ to have a chance for convergence at a reasonable rate. Furthermore if there are *several relatively close classes* that are to be separated via Kohonen network classification, one may never get there. If, however, a given class has *a wide spread of values*, several Kohonen neurons may be activated for the same class. Still, the latter situation can be subsequently corrected by the Grossberg layer which will then guide certain different Kohonen layer outputs to the same overall output.

The above considerations lead to a solution that *distributes the randomness of the* initial weights *to resemble the spread of the input vectors* of a given class.

To accomplish the latter *initialization strategy*, one may employ the *convex combination initialization method* as follows:

Set all initial weights to the *same value of* $1/\sqrt{N}$ where N is the number of inputs (dimension of $\mathbf{x}'$). Thus all input vectors will be of unity length (as required) since

$$N\left(\frac{1}{\sqrt{N}}\right)^2 = 1 \tag{8.13}$$

and *add* a small *noise ripple* component to these weights. Subsequently, set all x_i to satisfy

$$x_i^* = \gamma x_i + (1-\gamma)\frac{1}{\sqrt{N}} \tag{8.14}$$

with $\gamma \ll 1$ initially.

As the network trains, γ is gradually increased towards 1. Note that for $\gamma = 1$; $x_i^* = x_i$

Another approach is to *add noise* to the input vector. But this is slower than the earlier method.

A *third* alternative method starts with randomized normalized weights. But during the first few training sets all weights are adjusted, not *just those of the "winning neuron"*. Hence, the declaration of a "winner" will be delayed by a few iterations.

However, the best approach is to select a representative set of input vectors x and use these as initial weights s.t. each neuron will be initialized by one vector from that set.

8.4.3. *Interpolative mode layer*

Whereas a Kohonen layer retains only the "winning neuron" for a given class, the Interpolative Mode layer *retains a group of Kohonen neurons per a given class.* The retained neurons are those having the highest inputs. The number of neurons to be retained for a given class must be predetermined.

The outputs of that group will then be normalized to unit length. All other outputs will be zero.

8.5. Training of Grossberg Layers

A major asset of the Grossberg layer is the ease of its training. First the outputs of the Grossberg layer are calculated as in other networks, namely

$$g_i = \sum_j v_{ij} k_j = v_{ih} k_h = v_{ih} \tag{8.15}$$

K_j being the Kohonen layer outputs and v_{ij} denoting the Grossberg layer weights.

Obviously, only weights from non-zero Kohonen neurons (non-zero Grossberg layer inputs) are adjusted.

Weight adjustment follows the relations often used before, namely:

$$v_{ij}(n+1) = v_{ij}(n) + \beta [T_i - v_{ij}(n) k_j] \tag{8.16}$$

T_i being the *desired outputs* (*targets*), and for the $n+1$ iteration β being initially set to about 1 and is gradually reduced. Initially v_{ij} are randomly set to yield a vector of norm 1 per each neuron.

Hence, the weights will converge to the average value of the desired outputs to best match an input-output (x-T) pair.

8.6. The Combined Counter Propagation Network

We observe that the Grossberg layer is trained to converge to the desired (T) outputs whereas the Kohonen layer is trained to converge to the average inputs. Hence, the Kohonen layer is essentially a *pre-classifier* to *account for imperfect inputs*, the Kohonen layer being unsupervised while the Grossberg layer is supervised of.

If m target vectors T_j (of dimension p) are simultaneously applied at $m \times p$ outputs at the output side of the Grossberg layer to map Grossberg neurons then each set of p Grossberg neurons will converge to the appropriate target input, given the closest $\mathbf{x}$ input being applied at the Kohonen layer input at the time. The term Counter-Propagation (CP) is due to this application of input and target at each end of the network, respectively.

8.A. Counter Propagation Network Case Study*: Character Recognition

8.A.1. *Introduction*

This case study is concerned with recognizing three digits of "0", "1", "2" and "4". By using a Counter Propagation (CP) neural network. It involves designing the CP network, training it with standard data sets (8-by-8); testing the network using test data with 1, 5, 10, 20, 30, 40-bit errors and evaluating the recognition performance.

8.A.2. *Network structure*

The general CP structure is given in Fig. 8.A.1:

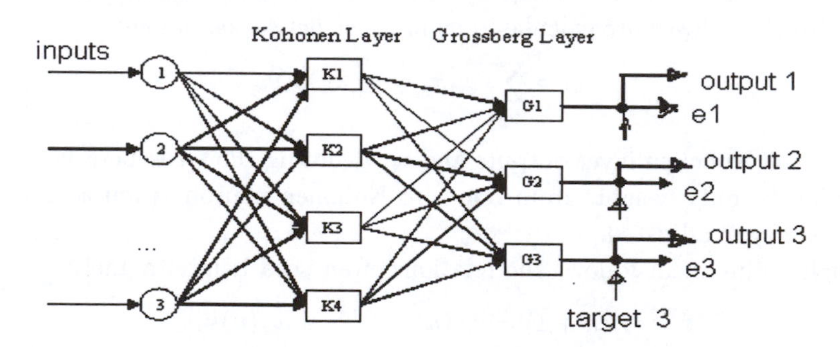

Fig. 8.A.1. Schematic diagram of the CP network.

A MATLAB-based design was established to create a default network:

Example: For creating a CP network with 64-input-neuron, 4-Kohonen-neuron, and 3-Grossberg-neuron:

```
%% Example:
%       neuronNumberVector = [64 4 3]
%% use:
%%       cp = createDefaultCP(neuronNumberVector);
%
```

*Computed by Yunde Zhong, ECE Dept., University of Illinois, Chicago, 2005.

8.A.3. *Network training*

(a) Training data set

The training data set applied to the CP network is as follows:

```
%%%%%%%%%%%%%%%%%%%%%%%%%%%%%
%% 2
%%%%%%%%%%%%%%%%%%%%%%%%%%%%%
classID = 0;
classID = classID + 1;
trainingData(1).input = [...
    -1 -1 -1 -1 -1 -1 -1 -1;...
    -1 -1 -1 -1 -1 -1 -1 -1;...
     1  1  1  1  1 -1 -1  1;...
     1  1  1  1 -1 -1  1  1;...
     1  1  1 -1 -1  1  1  1;...
     1  1 -1 -1  1  1  1  1;...
     1 -1 -1 -1 -1 -1 -1 -1;...
     1 -1 -1 -1 -1 -1 -1 -1 ...
  ];
trainingData(1).classID = classID;
trainingData(1).output = [0&1 0];
trainingData(1).name = '2';

%%%%%%%%%%%%%%%%%%%%%%%%%%%%%
%% 1
%%%%%%%%%%%%%%%%%%%%%%%%%%%%%
classID = classID + 1;
trainingData(2).input = [...
    1 1 1 -1 -1 1 1 1;...
    1 1 1 -1 -1 1 1 1;...
    1 1 1 -1 -1 1 1 1;...
    1 1 1 -1 -1 1 1 1;...
    1 1 1 -1 -1 1 1 1;...
    1 1 1 -1 -1 1 1 1;...
    1 1 1 -1 -1 1 1 1;...
    1 1 1 -1 -1 1 1 1 ...
  ];
trainingData(2).classID = classID;
trainingData(2).output = [0 0 1];
trainingData(2).name = '1';
```

```
%%%%%%%%%%%%%%%%%%%%%%%%%%%%%%
%% 4
%%%%%%%%%%%%%%%%%%%%%%%%%%%%%%
classID = classID + 1;
trainingData(3).input = [...
    -1 -1  1  1  1  1  1  1;...
    -1 -1  1  1  1  1  1  1;...
    -1 -1  1  1 -1 -1  1  1;...
    -1 -1  1  1 -1 -1  1  1;...
    -1 -1 -1 -1 -1 -1 -1 -1;...
    -1 -1 -1 -1 -1 -1 -1 -1;...
     1  1  1  1 -1 -1  1  1;...
     1  1  1  1 -1 -1  1  1 ...
  ];
trainingData(3).classID = classID;
trainingData(3).output = [1 0 0];
trainingData(3).name = '4';

%%%%%%%%%%%%%%%%%%%%%%%%%%%%%%
%% 0
%%%%%%%%%%%%%%%%%%%%%%%%%%%%%%
classID = classID + 1;
trainingData(4).input = [...
    1 -1 -1 -1 -1 -1 -1  1;...
    1 -1 -1 -1 -1 -1 -1  1;...
    1 -1 -1  1  1 -1 -1  1;...
    1 -1 -1  1  1 -1 -1  1;...
    1 -1 -1  1  1 -1 -1 -1;...
    1 -1 -1  1  1 -1 -1 -1;...
    1 -1 -1 -1 -1 -1 -1  1;...
    1 -1 -1 -1 -1 -1 -1  1 ...
  ];
trainingData(4).classID = classID;
trainingData(4).output = [0 0 0];
trainingData(4).name = '0';
```

(b) Setting of Weights:

(1) Get all training data vectors $X_i, i = 1, 2 \dots L$
(2) For each group of data vectors belonging to the same class, $X_i, i = 1, 2 \dots N$.
   (a) Normalize each X_i, $i = 1, 2 \dots N, X_i' = X_i / sqrt(\Sigma X^2 j)$
   (b) Compute the average vector $X = (\Sigma X_j')/N$
   (c) Normalize the average vector $X, X' = X/sqrt(X^2)$
   (d) Set the corresponding Kohonen Neuron's weights $W_k = X$
   (e) Set the Grossberg weights $[W_{1k} W_{1k} \dots W_{1k}]$ to the output vector Y
(3) Repeat step 2 until each class of training data is propagated into the network.

8.A.4. *Test mode*

The test data set is generated by a procedure, which adds a specified number of error bits to the original training data set. In this case study, a random procedure is used to implement this function.

Example:

testingData = getCPTestingData(trainingData, numberOfBitError, numberPerTrainingSet)

where the parameter, "numberOfBitError", is to specify the expected number of bit errors; "numberPerTrainingSet" is to specify the expected size of the testing data set. And the expected testing data set is gotten by the output parameter "testingData".

8.A.5. *Results and conclusions*

(a) Success rate vs. bit errors

In this experiment, a CP network with 64-input, 4-Kohonen-neuron, and 3-Grossberg-neuron is used. The success rate is tabulated as follows:

Success Rate		Number of Testing data set		
		12	100	1000
Number of Bit Error	1	100%	100%	100%
	5	100%	100%	100%
	10	100%	100%	100%
	20	100%	100%	100%
	30	100%	97%	98.2%
	40	91.6667%	88%	90.3%
	50	83.3333%	78%	74.9%

(b) Conclusions

(1) The CP network is robust and fast.
(2) CP network has high success rate even if in the case of large bit errors.

8.A.6. *Source codes (MATLAB)*

File #1

```
function cp= nnCP

%% Get the training data
[trainingData, classNumber] = getCPTrainingData;

%% Create a default CP network
outputLen = length(trainingData(1).output);
cp = createDefaultCP([64, classNumber, outputLen]);

%% Training the CP network
[cp] = trainingCP(cp, trainingData);

%% test the original training data set;
str = [];
tdSize = size(trainingData);
for n = 1: tdSize(2);
   [cp, output] = propagatingCP(cp, trainingData(n).input(:));

   [outputName, outputVector, outputError, outputClassID] = cpClassifier(cp,
trainingData);

   if strcmp(outputName, trainingData(n).name)
      astr = [num2str(n), '==>  Succeed!! The Error Is:', num2str(outputError)];
   else
      astr = [num2str(n), '==>  Failed!!'];
   end

   str = strvcat(str, astr);
end
output = str;
display(output);

%% test on the testing data set with bit errors
testingData = getCPTestingData(trainingData, 40, 250);
trdSize = size(trainingData);
tedSize = size(testingData);
str = [];
successNum = 0;
for n = 1: tedSize(2)
   [cp, output] = propagatingCP(cp, testingData(n).input(:));
   [outputName, outputVector, outputError, outputClassID] = cpClassifier(cp,
trainingData);

   strFormat = ' ';
   vStr = strvcat(strFormat,num2str(n));
   if strcmp(outputName, testingData(n).name)
      successNum = successNum + 1;
      astr = [vStr(2,:), '==> Succeed!! The Error Is:', num2str(outputError)];
   else
      astr = [vStr(2,:), '==> Failed!!'];
```

```
    end

    str = strvcat(str, astr);
end

astr = ['The success rate is:', num2str(successNum *100/tedSize(2)),'%'];
str = strvcat(str, astr);
testResults = str;
display(testResults);
```

File #2

```
%%%%%%%%%%%%%%%%%%%%%%%%%%%%%%%%%%%%%%%%%%%%%%%%%%%%%%%%%%%%%%%%%%%%%%%%%%%%%%%%%%%%%%%%%%%%
%% A function to create a default Counter Propagation model
%%
%% input parameters:
%% neuronNumberVector to specify neuron number in each layer
%%
%% Example #1:
%%          neuronNumberVector = [64 3 3]
%%    use:
%%          cp = createDefaultCP(neuronNumberVector);
%%
%% Author: Yunde Zhong
%%%%%%%%%%%%%%%%%%%%%%%%%%%%%%%%%%%%%%%%%%%%%%%%%%%%%%%%%%%%%%%%%%%%%%%%%%%%%%%%%%%%%%%%%%%%%

function cp = createDefaultCP(neuronNumberVector)

cp = [];

if nargin < 1
   display('createDefaultCP.m needs one parameter');
   return;
end

nSize = length(neuronNumberVector);
if nSize ~= 3
   display('error parameter when calling createDefaultCP.m');
   return;
end

%% nn network paramters
cp.layerMatrix = neuronNumberVector;

%% Kohonen layer
aLayer.number = neuronNumberVector(2);
aLayer.error = [];
aLayer.output = [];
aLayer.neurons = [];
aLayer.analogOutput = [];
aLayer.weights = [];

for ind = 1: aLayer.number
```

```
    %% create a default neuron
    inputsNumber = neuronNumberVector(1);
    weights = ones(1,inputsNumber) / sqrt(aLayer.number);

    aNeuron.weights = weights;
    aNeuron.weightsUpdateNumber = 0;
    aNeuron.z = 0;
    aNeuron.y = 0;

    aLayer.neurons = [aLayer.neurons, aNeuron];
    aLayer.weights = [aLayer.weights; weights];
end

cp.kohonen = aLayer;

%% Grossberg Layer
aLayer.number = neuronNumberVector(3);
aLayer.error = [];
aLayer.output = [];
aLayer.neurons = [];
aLayer.analogOutput = [];
aLayer.weights = [];

%% create a default layer
for ind = 1: aLayer.number
    %% create a default neuron
    inputsNumber = neuronNumberVector(2);

    weights = zeros(1,inputsNumber);
    aNeuron.weights = weights;
    aNeuron.weightsUpdateNumber = 0;
    aNeuron.z = 0;
    aNeuron.y = 0;

    aLayer.neurons = [aLayer.neurons, aNeuron];
    aLayer.weights = [aLayer.weights; weights];
end

cp.grossberg = aLayer;
```

File #3

```
function [trainingData, classNumber] = getCPTrainingData

trainingData = [];
classNumber =[];

classID = 0;


%%%%%%%%%%%%%%%%%%%%%%%%%%%%%%
%% 2
%%%%%%%%%%%%%%%%%%%%%%%%%%%%%%
classID = classID + 1;
trainingData(1).input = [...
      -1 -1 -1 -1 -1 -1 -1 -1;...
      -1 -1 -1 -1 -1 -1 -1 -1;...
       1  1  1  1  1 -1 -1  1;...
       1  1  1  1 -1 -1  1  1;...
       1  1  1 -1 -1  1  1  1;...
       1  1 -1 -1  1  1  1  1;...
       1 -1 -1 -1 -1 -1 -1 -1;...
       1 -1 -1 -1 -1 -1 -1 -1
   ];
trainingData(1).classID = classID;
trainingData(1).output = [0&1 0];
trainingData(1).name = '2';


%%%%%%%%%%%%%%%%%%%%%%%%%%%%%%
%% 1
%%%%%%%%%%%%%%%%%%%%%%%%%%%%%%
classID = classID + 1;
trainingData(2).input = [...
      1 1 1 -1 -1 1 1 1;...
      1 1 1 -1 -1 1 1 1;...
      1 1 1 -1 -1 1 1 1;...
      1 1 1 -1 -1 1 1 1;...
      1 1 1 -1 -1 1 1 1;...
      1 1 1 -1 -1 1 1 1;...
      1 1 1 -1 -1 1 1 1;...
      1 1 1 -1 -1 1 1 1
   ];
trainingData(2).classID = classID;
trainingData(2).output = [0 0 1];
trainingData(2).name = '1';
```

```
%%%%%%%%%%%%%%%%%%%%%%%%%%%%%%
%% 4
%%%%%%%%%%%%%%%%%%%%%%%%%%%%%%
classID = classID + 1;
trainingData(3).input = [...
     -1 -1  1  1  1  1  1  1;...
     -1 -1  1  1  1  1  1  1;...
     -1 -1  1  1 -1 -1  1  1;...
     -1 -1  1  1 -1 -1  1  1;...
     -1 -1 -1 -1 -1 -1 -1 -1;...
     -1 -1 -1 -1 -1 -1 -1 -1;...
      1  1  1  1 -1 -1  1  1;...
      1  1  1  1 -1 -1  1  1 ...
  ];
trainingData(3).classID = classID;
trainingData(3).output = [1 0 0];
trainingData(3).name = '4';

%%%%%%%%%%%%%%%%%%%%%%%%%%%%%%
%% 0
%%%%%%%%%%%%%%%%%%%%%%%%%%%%%%
classID = classID + 1;
trainingData(4).input = [...
     1 -1 -1 -1 -1 -1 -1  1;...
     1 -1 -1 -1 -1 -1 -1  1;...
     1 -1 -1  1  1 -1 -1  1;...
     1 -1 -1  1  1 -1 -1  1;...
     1 -1 -1  1  1 -1 -1 -1;...
     1 -1 -1  1  1 -1 -1 -1;...
     1 -1 -1 -1 -1 -1 -1  1;...
     1 -1 -1 -1 -1 -1 -1  1 ...
  ];
trainingData(4).classID = classID;
trainingData(4).output = [0 0 0];
trainingData(4).name = '0';

%% Other parameters
classNumber = classID;
```

File #4

```
function cp = trainingCP(cp, trainingData )

if nargin < 2
   display('trainingCP.m needs at least two parameter');
   return;
end

datbasetSize = size(trainingData);
kWeights = [];
gWeights = zeros(cp.grossberg.number,cp.kohonen.number);
for datbasetIndex = 1: datbasetSize(2)
   mIn = trainingData(datbasetIndex).input(:);
   mOut = trainingData(datbasetIndex).output(:);
   mClassID = trainingData(datbasetIndex).classID;

   mIn = mIn / sqrt(sum(mIn.*mIn));

   %% training the Kohonen Layer
   oldweights = cp.kohonen.neurons(mClassID).weights;
   weightUpdateNumber = cp.kohonen.neurons(mClassID).weightsUpdateNumber + 1;

   if weightUpdateNumber >&1
      mIn = (oldweights * weightUpdateNumber + mIn) / weightUpdateNumber;
      mIn = mIn / sqrt(sum(mIn .* mIn));
   end

   cp.kohonen.neurons(mClassID).weights = mIn';
   cp.kohonen.neurons(mClassID).weightsUpdateNumber = weightUpdateNumber;

   kWeights = [kWeights; mIn'];

   %% training the Grossberg Layer
   if weightUpdateNumber >&1
      mOut = (mOut * weightUpdateNumber + mOut) / weightUpdateNumber;
   end

   gWeights(:,mClassID) = mOut;
end

for gInd = 1: cp.grossberg.number
   cp.grossberg.neurons(gInd).weights = gWeights(gInd,:);
end

cp.kohonen.weights = kWeights;
cp.grossberg.weights = gWeights;
```

File #5

```
function [cp, output] = propagatingCP(cp, inputData)

output = [];
if nargin < 2
   display('propagatingCP.m needs two parameters');
   return;
end


% propagation of Kohonen Layer
zOut = cp.kohonen.weights * inputData;
[zMax, zMaxInd] = max(zOut);
yOut = zeros(size(zOut));
yOut(zMaxInd) = 1;

cp.kohonen.analogOutput = zOut;
cp.kohonen.output = yOut;

for kInd =&1 : cp.kohonen.number
   cp.kohonen.neurons(kInd).z = zOut(kInd);
   cp.kohonen.neurons(kInd).y = yOut(kInd);
end

% propagation of Grossberg Layer
zOut = cp.grossberg.weights * yOut;
yOut = zOut;
cp.grossberg.analogOutput = zOut;
cp.grossberg.output = yOut;

for gInd =&1 : cp.grossberg.number
   cp.grossberg.neurons(gInd).z = zOut(gInd);
   cp.grossberg.neurons(gInd).y = yOut(gInd);
end
```

File #6

```
function [outputName, outputVector, outputError, outputClassID] = cpClassifier(cp, trainingData)

outputName = [];
outputVector = [];

if nargin < 2
   display('cpClassifier.m needs at least two parameter');
   return;
end

dError = [];
dataSize = size(trainingData);
output = cp.grossberg.output;
for dataInd =1 : dataSize(2)
   aSet = trainingData(dataInd).output(:);
```

```
    vDiff = abs(aSet - output);
    vDiff = vDiff.*vDiff;

    newError = sum(vDiff);

    dError = [dError, newError];
end

if ~isempty(dError)
    [eMin, eInd] = min(dError);
    outputName = trainingData(eInd).name;
    outputVector = trainingData(eInd).output;
    outputError = eMin;
    outputClassID = trainingData(eInd).classID;
end
```

File #7

```
function testingData = getCPTestingData(trainingData, numberOfBitError,
numberPerTrainingSet)

testingData = [];

tdSize = size(trainingData);
tdSize = tdSize(2);

ind = 1;
for tdIndex = 1: tdSize
    input = trainingData(tdIndex).input;
    name = trainingData(tdIndex).name;
    output = trainingData(tdIndex).output;
    classID = trainingData(tdIndex).classID;
    inputSize = size(input);

    for ii = 1: numberPerTrainingSet
        rowInd = [];
        colInd = [];

        flag = ones(size(input));
        bitErrorNum = 0;
        while bitErrorNum < numberOfBitError
            x = ceil(rand(1) * inputSize(1));
            y = ceil(rand(1) * inputSize(2));
            if x <= 0
                x = 1;
            end
            if y <= 0
                y = 1;
            end
            if flag(x, y) ~= &-1
                bitErrorNum = bitErrorNum + 1;
                flag(x, y) == -1;
                rowInd = [rowInd, x];
```

```
            colInd = [colInd, y];
         end
      end

      newInput = input;

      for en = 1:numberOfBitError
         newInput(rowInd(en), colInd(en)) = newInput(rowInd(en), colInd(en)) * (-1);
      end

      testingData(ind).input = newInput;
      testingData(ind).name = name;
      testingData(ind).output = output;
      testingData(ind).classID = classID;

      ind = ind + 1;
   end
end
```

Chapter 9

Adaptive Resonance Theory

9.1. Motivation

The Adaptive Resonance Theory (ART) was originated by Carpenter and Grossberg (1987a) for the purpose of developing artificial neural networks whose manner of performance, especially (but not only) in pattern recognition or classification tasks, is closer to that of the biological neural network (NN) than was the case in the previously discussed networks. One of their main goals was to come up with neural networks that can preserve the biological network's plasticity in learning or in recognizing new patterns, namely, in learning without having to erase (forget) or to substantially erase earlier learned patterns.

Since the purpose of the ART neural network is to closely approximate the biological NN, the ART neural network needs no "teacher" but functions as an unsupervised self-organizing network. Its ART-I version deals with binary inputs. The extension of ART-I known as ART-II [Carpenter and Grossberg, 1987b] deals with both analog patterns and with patterns represented by different levels of grey.

9.2. The ART Network Structure

The ART network consists of 2 layers; (I) a Comparison Layer (CL) and (II) a Recognition Layer (RL), which are interconnected. In addition, the network consists of two Gain elements, one, (G_1) feeding its output g_1 to the Comparison layer and the second, (G_2) feeding its output g_2 to the Recognition Layer, and thirdly, a Reset element where the comparison, as performed in the Comparison Layer, is evaluated with respect to a preselected tolerance value ("vigilance" value). See Fig. 9.1.

9.2. (a) *The comparison layer (CL)*

A binary element x_j of the m-dimensional input vector $\mathbf{x}$ is inputted into the jth ($j = 1 \cdots m$; $m = \dim(\mathbf{x})$) neuron of the CL. The jth neuron is also inputted by a weighted sum (p_j) of the recognition-output vector $\mathbf{r}$ from the RL where

$$p_j = \sum_{i=1}^{m} t_{ij} r_i \tag{9.1}$$

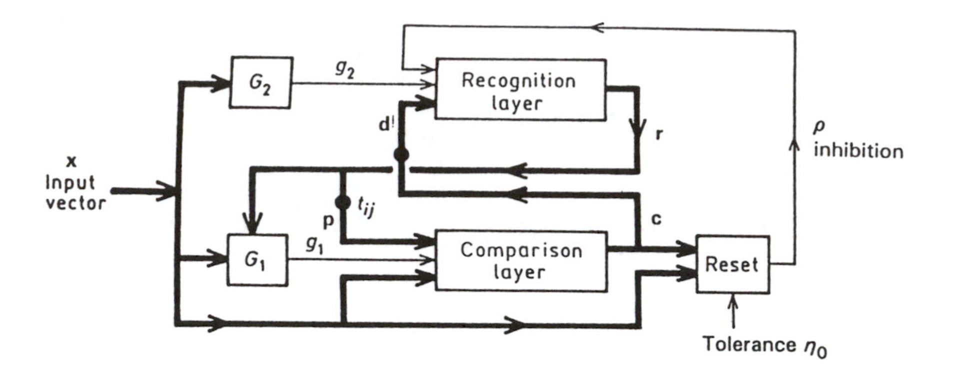

Fig. 9.1. ART-I network schematic.

r_i being the ith component of the m-dimensional recognition-output vector $\mathbf{r}$ of the RL layer and n being the number of categories to be recognized.

Furthermore, all CL neurons receive the same g_1 scalar output of the same element G_1. The m-dimensional ($m = \dim(\mathbf{x})$) binary comparison-layer output vector $\mathbf{c}$ of the CL layer initially equals the input vector, namely, at the initial iteration

$$c_j(0) = x_j(0) \tag{9.2}$$

Also, initially:

$$g_1(0) = 1 \tag{9.3}$$

The CL's output vector $\mathbf{c}$ satisfies a (two-thirds) majority rule requirement s.t. its output is

$$c_j = 1$$

only if at least two of this (CL) neuron's three inputs are 1. Hence, Eqs. (9.2), (9.3) imply, by the "two-thirds majority" rule, that initially

$$\mathbf{c}(0) = \mathbf{x}(0) \tag{9.4}$$

since initially no feedback exists from the *RL* layer, while $g_1(0) = 1$.

9.2. (b) *The recognition layer (RL)*

The RL layer serves as a classification layer. It receives as its input an n-dimensional weight vector $\mathbf{d}$ with elements d_j, which is the weighted form of the CL's output vector c; s.t.

$$d_j = \sum_{i=1}^{m} b_{ji} c_i = \mathbf{b}_j^T \mathbf{c} \; ; \quad \mathbf{b}_j \triangleq \begin{bmatrix} b_{j1} \\ \vdots \\ b_{jm} \end{bmatrix} ; \qquad \begin{aligned} i &= 1, 2, \ldots m \, ; \\ j &= 1, 2, \ldots n \, ; \\ m &= \dim(x) \\ n &= \text{number of categories} \end{aligned} \tag{9.5}$$

where b_{ji} are real numbers.

The RL neuron with the maximal (winning) d_j *will output a* "1" as long as $g_2 = 1$. All others will output a zero. Hence, the RL layer serves to classify its input vector. The weights b_{ij} of the jth (winning) RL neuron that fires (having maximal output d_j) constitute an exemplar of the pattern of vector $\mathbf{c}$, in similarity to the properties of the BAM memory discussed earlier (Sec. 7.3), noting that an output d_j satisfies

$$d_j = \mathbf{c}^T\mathbf{c} \quad \text{at maximum (as in the Kohonen layer)} \tag{9.6}$$

d_j being the maximal possible outcome of Eq. (9.5), since $\mathbf{b}_j = \mathbf{c}$; $\mathbf{d}_{i\neq j} = 0$.

We achieve the *locking of one neuron* (the winning neuron) to the *maximal output* by outputting a winner-take-all (as in Sec. 8.2):

$$r_j = 1 \tag{9.7}$$

while all other neurons yield

$$r_{i\neq j} = 0 \quad \text{if} \quad \rho = 0 \text{ (no inhibition)} \tag{9.8}$$

For this purpose an interconnection scheme is employed in the RL that is based on *lateral inhibition*. The lateral inhibition interconnection is as in Fig. 9.2, where the output r_i of each neuron (i) is connected via an *inhibitory* (negative) weight matrix $L = \{l_{ij}\}$, $i \neq j$, where $l_{ij} < 0$ to any other neuron (j). Hence, a neuron with a

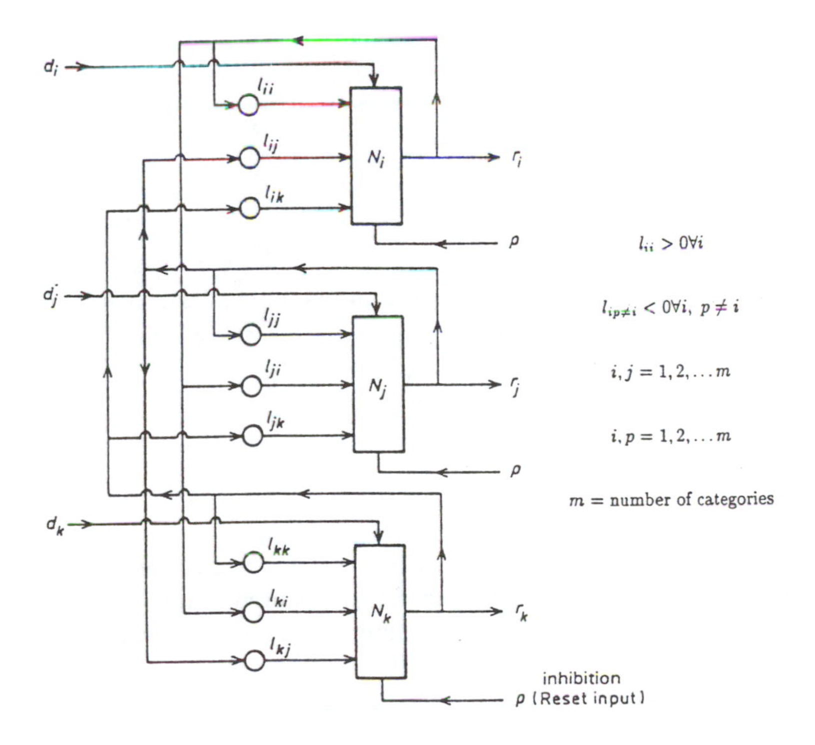

Fig. 9.2. Lateral inhibition in an RL layer of an ART-I network.

large output inhibits all other neurons. Furthermore, positive feedback $l_{jj} > 0$ is employed such that each neuron's output r_j is fed back with a positive weight to its own input to reinforce its output if it is to fire (to output a "one"). This *positive reinforcement* is termed as *adaptive resonance* to motivate the resonance term in "ART".

9.2. (c) *Gain and reset elements*

The gain elements feed the *same* scalar output to all neurons concerned as in Fig. 1, g_1 being inputted to the CL neurons and g_2 to RL neurons, where:

$$\begin{aligned} g_2 &= OR(\mathbf{x}) = OR(x_1 \cdots x_N) \\ g_1 &= \overline{OR(\mathbf{r})} \cap OR(\mathbf{x}) \\ &= \overline{OR(r_1 \cdots r_N)} \cap OR(x_1 \cdots x_N) = g_2 \cap \overline{OR(\mathbf{r})} \end{aligned} \tag{9.9}$$

Hence, if at least one element of $\mathbf{x}$ is 1 then $g_2 = 1$. Also, if any element of $g_2 = 1$ but also no elements of $\mathbf{r}$ is 1 then $g_1 = 1$, else $g_1 = 0$. (See Table 9.1). Note that the overhead bar denotes negation, whereas $\cap$ denotes a logical "and" (intersection). Also, note that if $OR(\mathbf{x})$ is zero, then $OR(\mathbf{r})$ is always zero, by the derivation of $\mathbf{r}$ as above.

Table 9.1.

OR(**x**)	OR(**r**)	$\overline{\text{OR}(\mathbf{r})}$	g_1
0	0	1	0
1	0	1	1
1	1	0	0
0	1	0	0

Finally, the *reset element* evaluates the degree of similarity between the input vector $\mathbf{x}$ and the CL output vector $\mathbf{c}$ in terms of the ratio η, where:

$$\eta = \frac{\text{No. of "1"s in } \mathbf{c}}{\text{No. of "1"s in } \mathbf{x}} \tag{9.10}$$

Subsequently, if

$$\eta < \eta_0 \tag{9.11}$$

η_0 being a pre-set initial *tolerance* (vigilance) value, then a reset signal (ρ) is outputted to inhibit the particular RL neuron that has fired at the given iteration. See Fig. 9.2. A reset factor based of the Hamming distance between vectors $\mathbf{c}$ and $\mathbf{x}$ can also be considered.

9.3. Setting-Up of the ART Network

9.3. (a) *Initialization of weights*

The CL weight matrix B is initialized [Carpenter and Grossberg, 1987a] as follows:

$$b_{ij} < \frac{E}{E+m-1} \quad \forall i,j \tag{9.12}$$

where

$$m = \dim(\mathbf{x})$$

$$E > 1 \text{ (typically } E = 2\text{)}$$

The RL weight matrix T is initialized such that

$$t_{ij} = 1 \quad \forall i,j \tag{9.13}$$

(See: Carpenter and Grossberg, 1987a)
The tolerance level η_0 (vigilance) is chosen as

$$0 < \eta_0 < 1 \tag{9.14}$$

A high η_0 yields fine discrimination whereas a low η_0 allows grouping of more dissimilar patterns. Hence, one may start with lower η_0 and raise it gradually.

9.3. (b) *Training*

The training involves the setting of the weight matrices B (of the RL) and T (of the CL) of the ART network.

Specifically, the network may first be exposed for brief periods to successive input vectors, not having time to converge to any input vector but only to approach some settings corresponding to some averaged **x**.

The parameters b_{ij} of vector $\mathbf{b}_j$ of B are set according to:

$$b_{ij} = \frac{Ec_i}{E+1+\sum_k c_k} \tag{9.15}$$

where $E > 1$ (usually, $E = 2$)
c_i = the ith component of vector **c** where j corresponds to the *winning neuron*.

Furthermore, the parameter t_{ij} of T are set such that

$$t_{ij} = c_i \quad \forall i = 1 \cdots m; \quad m = \dim(\mathbf{x}) \tag{9.16}$$

j denoting the *winning neuron*.

9.4. Network Operation

(a) Initially, at iteration 0, $\mathbf{x} = \mathbf{0}$. Hence, by Eq. (9.9),

$$g_2(0) = 0$$

and

$$g_1(0) = 0$$

Consequently $\mathbf{c}(0) = \mathbf{0}$ by Eq. (4).
Also, since $g_2(0) = 0$, the output vector $\mathbf{r}$ to the CL is, by the majority $(\frac{2}{3})$ rule that governs both layers is: $\mathbf{r}(0) = \mathbf{0}$.
(b) Subsequently, when a vector $\mathbf{x} \neq \mathbf{0}$ is being applied, no neuron has an advantage on any other. Since now $\mathbf{x} \neq \mathbf{0}$, then $g_2 = 1$ and thus also $g_1 = 1$ (due to $\mathbf{r}(0) = \mathbf{0}$). Hence

$$\mathbf{c} = \mathbf{x} \tag{9.17}$$

by the majority rule described earlier.
(c) Subsequently, by Eqs. (9.5), (9.6) and the properties of the RL, the jth of the RL neuron, which best matches vector $\mathbf{c}$ will be the only RL neuron to fire (to output a one). Hence $r_j = 1$ and $r_{l \neq j} = 0$ to determine vector $\mathbf{r}$ at the output of the RL. Note that if several neurons have same d then the first one will be chosen (lowest j).

Now $\mathbf{r}$ as above is fed back to the CL such that it is inputted to the CL neurons via weights t_{ij}. The m-dimensional weight vector $\mathbf{p}$ at the input of the CL thus satisfies namely

$$\mathbf{p}_j = \mathbf{t}_j; \quad \mathbf{t}_j \text{ denoting a vector of } T \tag{9.18a}$$

for winning neuron, and

$$\mathbf{p}_j = \mathbf{0} \tag{9.18b}$$

otherwise
Notice that $r_j = 1$ and t_{ij} are binary values. The values t_{ij} of T of the CL are set by the training algorithm to correspond to the real weight matrix B (with elements b_{ij}) of the RL.

Since now $\mathbf{r} \neq \mathbf{0}$, then g_1 becomes 0 by Eq. (9.9) and by the majority rule, the CL neurons which receive that non-zero components of $\mathbf{x}$ and of $\mathbf{p}$ will fire (to output a "one" in the CL's output vector $\mathbf{c}$). Hence, the outputs of the RL force these components of $\mathbf{c}$ to zero where $\mathbf{x}$ and $\mathbf{p}$ do not have matching "ones".
(d) If classification is considered by the reset element to be adequate, then classification is stopped. Go to (f), else: A considerable mismatch between vectors $\mathbf{p}$ and $\mathbf{x}$ will result in a considerable mismatch (in terms of "one's") between vectors $\mathbf{x}$ and $\mathbf{c}$. This will lead to a low η as in Eq. (9.10) as computed by the *reset element* of the network such that $\eta < \eta_0$. This, in turn, *will inhibit* the firing neuron of the

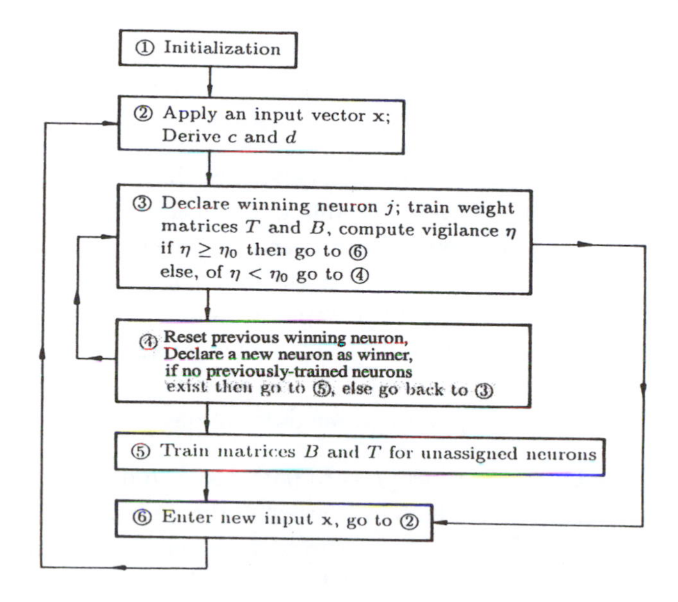

Fig. 9.3. Flow-chart of ART-I operation.

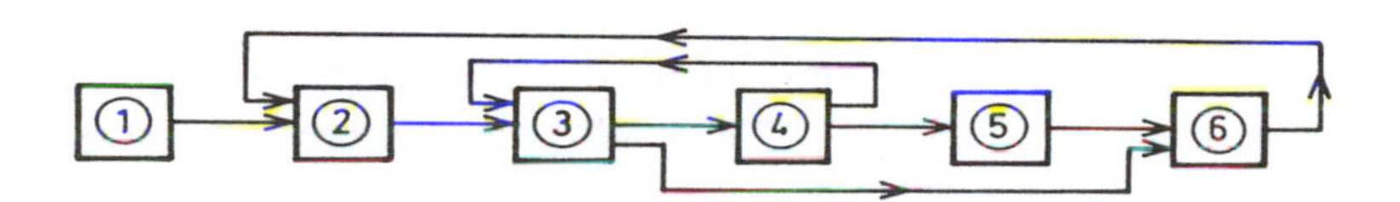

Fig. 9.4. Simplified ART-I flow chart (encircled numbers: as in Fig. 9.3).

RL. Since now $\eta < \eta_0$, then also all elements of $\mathbf{r} = \mathbf{0}$. Hence, $g_1 = 1$ and $\mathbf{x} = \mathbf{c}$ by the majority rule. Consequently, as long as neurons that are weighted still exist, a *different neuron* in the RL will win (the last winner being now inhibited), go to (c). (If at this iteration the reset element still considers the fit (classification) to be inadequate, the cycle is repeated. Now *either* a match will eventually be found: In that case the network will *enter a training* cycle where the weights vectors t_j and b_j associated with the firing RL neuron are modified to match the input x considered.) *Alternatively*, if no neuron matches the input within the tolerance, then go to (e).
(e) Now, a previously unassigned neuron is assigned weights vectors $\mathbf{t}_j$, and $\mathbf{b}_j$ to match the input vector $\mathbf{x}$. In this manner, the network does not lose (forget) previously learned patterns, but is also able to learn new ones, as does a biological network
(f) Apply new input vector.

The procedure above is summarized in Fig. 9.3. The categories (classes, patterns) which have been trained to be recognized are thus in terms of vectors (columns $\mathbf{t}_j$) of the weights matrix T; j denoting the particular category (class) considered, and $j = 1, 2, \ldots, n$, where n is the total number of classes considered.

9.5. Properties of ART

One can show [Carpenter and Grossberg, 1987a] that the ART network has several features that characterize the network, as follows:

1. Once the network stabilizes (weights reach steady state), the application of an input vector x, that has been used in training, will activate the correct RL neuron without any search (iterations). This property of *direct access* is similar to the rapid retrieval of previously learned patterns in biological networks.
2. The search process stabilizes at the *winning* neuron.
3. The training is *stable* and does not switch once a winning neuron has been identified.
4. The training stabilizes in a *finite number* of iterations.

To proceed from binary patterns (0/1) to patterns with different shades of gray, the authors of the ART-I network above developed the ART-II network [Carpenter and Grossberg, 1987b] which is not discussed here but which follows the basic philosophy of the ART-I network described above while extending it to continuous inputs.

The above indicates that the ART network; have many desirable features of biological networks, such as its being unsupervised, plastic, stable and of a finite number of iterations, and having immediate recall of previously learned patterns. The *main shortcoming* of the ART-I network is that a missing neuron destroys the whole learning process (since $\mathbf{x}$ and $\mathbf{c}$ must be of the same dimension). This *contrasts* an important property of biological neural networks. Whereas many of the properties of the ART network outlined above were missing in the previous networks, the latter shortcoming is not fundamental to the networks of the earlier chapters. It also leads us to consider the neural network designs of the next chapters, specifically, the *Cognitron/Neocognitron* neural network design and the LAMSTAR network design, which (among other things) avoid the above shortcoming.

9.6. Discussion and General Comments on ART-I and ART-II

We observe that the ART-I network incorporates the best features of practically all previously discussed neural networks. It employs a multilayer structure. It utilizes feedback as does the Hopfield network, though in a different form. It employs BAM learning as in the Hopfield network (Sec. 7.3) or as in the Kohonen layers discussed in counterpropagation designs (Chap. 6). It also uses a "winner-take-all" rule and as do the Kohonen (SOM) layer. In contrast to these other networks, it however incorporates all and not just some of these features in one design, noting that essentially all these features are also found in biological neural networks. In addition, and again in similarity to biological networks, it employs inhibition and via its reset function it has a plasticity feature. The ART-network's shortcomings in its failure to perform when one or more neurons are missing or malfunctioning, can be overcome by a modified design, such as proposed by Graupe and Kordylewski, (1995), where it is also shown how ART-I is modified to use non-binary inputs through performing simple input-coding. In general, ART-II networks are specifically derived for continuous (analog) inputs and their structure is thus modified to allow them to employ such inputs, whereas the network of the case study below still employs the ART-I architecture and its main modifications relative to the standard ART-I, if there is such a standard, as is necessitated for the specific application. Indeed, many applications do require modifications from a standard network whichever it is, for best results. Hence the modification below is not an exception.

9.A. ART-I Network Case Study*: Character Recognition

9.A.1. *Introduction*

This case study aims at implementing simple character recognition using the ART-I neural network.

The ART network consists of 2 layers: a comparison Layer and a Recognition Layer.

The general structure of the ART-I network is given in Fig. 9.A.1:

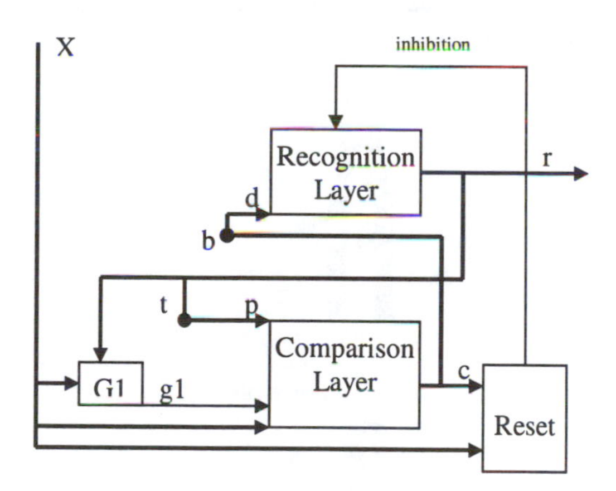

Fig. 9.A.1. General schematic diagram of the ART-I neural network.

The network's design follows Sect. 9.2 above.

9.A.2. *The data set*

Our Artificial neural network must be recognizing some characters in a 6×6 grid. This Neural network is tested on the following 3 characters:

*Computed by Michele Panzeri, ECE Dept., University of Illinois, Chicago, 2005.

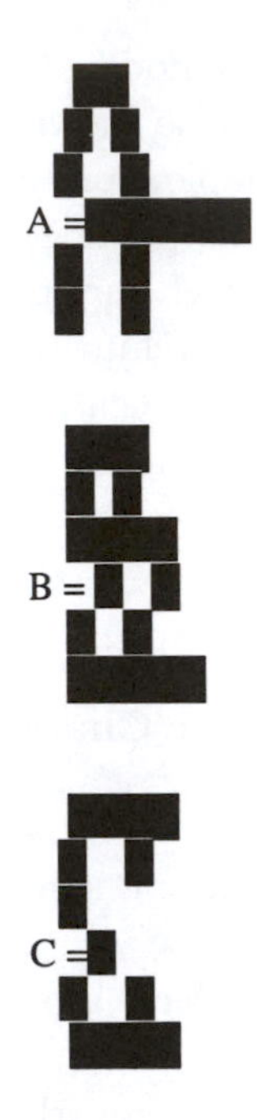

Moreover the network is tested on character with some noise (from 1 to 14 bits of error), as in the following examples:

1 bit of noise on X0:

5 bits of noise on X0:

Also, the network must be able to understand if a character does not belong to the predefined set. For example the following characters are not in the predefined (trained) set:

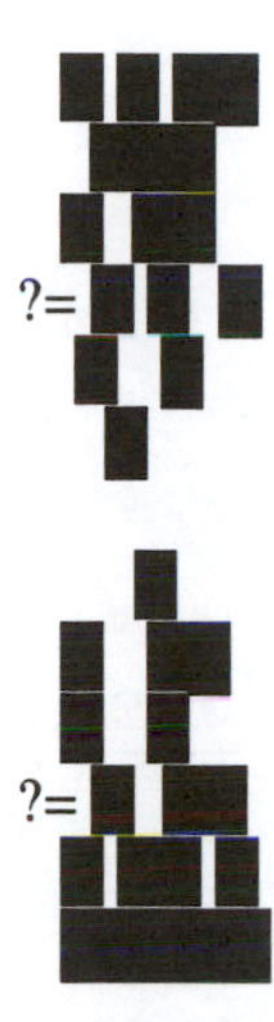

We could consider a large number of characters not predefined, for this reason these character are simply created randomly.

9.A.3. *Network design*

(a) Network structure

To solve this problem the network structure of Fig. 9.A.2 is adopted, where $x0 \cdots x35$ is the array that implements the 6×6 grid in input to our network.

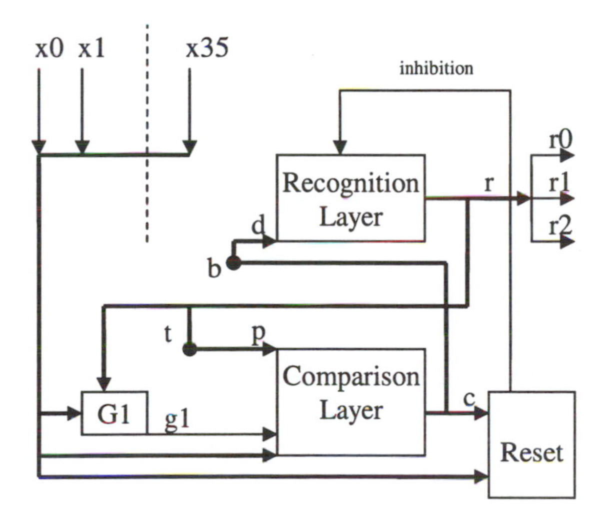

Fig. 9.A.2: The ART network for the present study.

(b) Setting of weights

The weights are initially set with:

$$b_{ij} = \frac{E}{E + m - 1}$$

and

$$t_{ij} = 1$$

During the training phase we update the weights according to:

$$b_{ij} = \frac{E * c_i}{E + 1 + \sum_k c_k}$$

and

$$t_{ij} = c_i$$

where j is the winning neuron

(c) Algorithm basics

The following is the procedure for the computation of the ART network:

(1) Assign weights as explained before.
(2) Train the network with the formulas explained before with some character.

Now we can test the network with pattern with noise and test the network with pattern that does not belong to the original set.

To distinguish the known pattern from the unknown pattern the network compute the following

$$\eta = \frac{\#1\ in\ c}{\#1\ in\ x}$$

If $\eta < \eta_0$ than the Recognition Layer is inhibited and all his neuron will output "0".

We comment that while the above setting of (is very simple (and satisfactory in our simple application, it is often advisable to set (by using a Hamming distance (see Sect. 7.3 above).

(d) Network training

This network is trained as follows (code in Java):

```
for(int i=0;i<m;i++){
        for(int j=0;j£<£nx;j++){
                b[j][i]=E/(E+m-1);
                t[i][j]=1;
        }
```

```
}

a();

r[0]=1;
r[1]=0;
r[2]=0;
compute();

sumck=0;
for(int k=0;k<m;k++){
        sumck+=x[k];
}

for(int i=0;i<m;i++){
        b[0][i]=E*x[i]/(E+1+sumck);
        t[i][0]=x[i];
}

[...]//The same for b and c

for(int i=0;i<m;i++){
        c[i]=0;
}

for(int j=0;j<nx;j++){
        r[j]=0;
}
```

This is the code used for the evaluation of the network:

```
int sumc=0;
int sumx=0;
for(int i=0;i<m;i++){

        p[i]=0;
        for(int j=0;j<nx;j++){
                p[i]+=t[i][j]*r[j];
        }

        if(p[i]>0.5){
                p[i]=1;

        }else{
                p[i]=0;
        }
        if(g1){
                c[i]=x[i];
        }else{
                if((p[i]+x[i])>=2.0){
```

```
                        c[i]=1;

                }else{
                        c[i]=0;
                }
        }
        if(c[i]==1){
                sumc++;
        }

        if(x[i]==1){
                sumx++;
        }
}

if((((double)sumc)/((double)sumx))<pho0){

        for(int j=0;j<nx;j++){
                 r[j]=0;
        }

}else{
        double max=Double.MIN_VALUE;
        int rmax=-1;

        for(int i=0;i<nx;i++){

                d[i]=0;
                for(int j=0;j<m;j++){
                         d[i]+=b[i][j]*c[j];
                }
        }

        for(int i=0;i<nx;i++){
                if(d[i]>0.5 &&d[i]>max){
                        max=d[i];
                        rmax=i;
                }
        }

        for(int i=0;i<nx;i++){
                if(i==rmax){
                       r[i]=1;

                }else{
                       r[i]=0;

                }
        }
}
```

The full code of the implementation is on the appendix.

9.A.4. *Performance results and conclusions*

The network is simulated to investigate how robust it is.

For this reason we simulated this network adding 1 to 18 bits of noise. The results of this simulation are collected in the following table.

The first column contains the number of bits of noise added at the input and the second column gives the percentage of error.

Number of bit of noise	Error %
0	0
1	0
2	0
3	0
4	0
5	0
6	0
7	0
8	0
9	5,8
10	7,5
11	18,4
12	21,9
13	34,1
14	35,1
15	46,2
16	49,4
17	56
18	63,4

Figure 9.A.3. provides a graphical display of the trend of the recognition error.

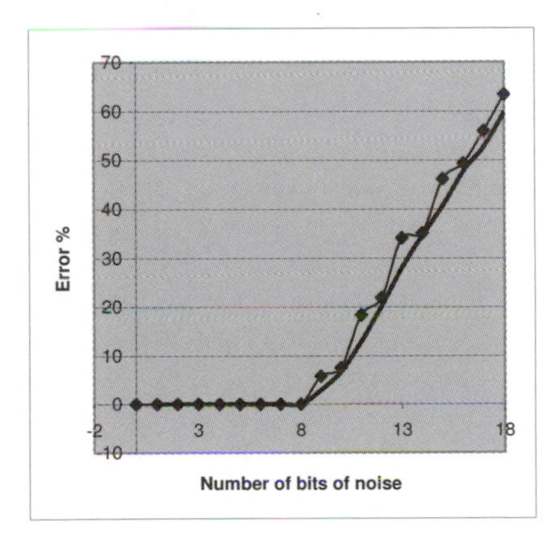

Fig. 9.A.3. Error percentage vs. number of error bits.

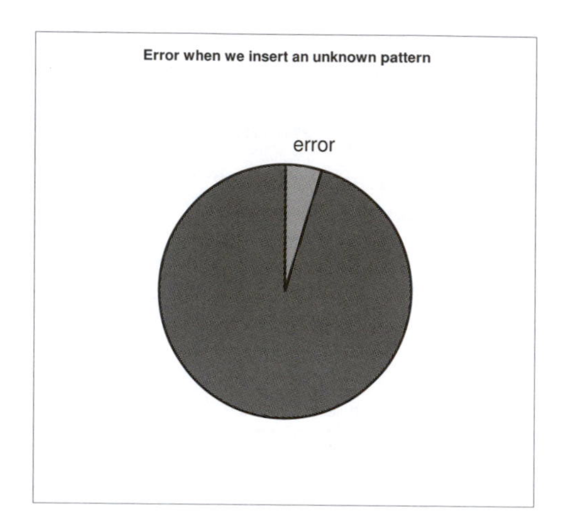

Fig. 9.A.4. Recognition error when dataset includes an untrained character.

As can be seen from the table and from Fig. 9.A.3, the network is incredibly robust at the noise, With noise of 8 bits or less the network always recognizes the correct pattern. With noise of 10 bits (of a total of 36 bits), the networks identifies correctly in the 90% of the cases.

We also investigated the behavior of the network when we use an unknown (untrained) character. For this reason we did another simulation to test if the network activates the output called "No pattern" when presented with an untrained character.

In this case the network still performs well (it understands that this is not a usual pattern) at a success rate of 95.30% (i.e. it fails in 4,70% of the cases). See Fig. 9.A.4.

9.A.5. *Code for ART neural network (Java)*

```
public class Network {

        //Number of inputs 6x6
        final int m=36;
        //Number of char/Neuron for each layer
        final int nx=3;

        final double pho0=0.65;
        final int E=2;

        //State of the net
        public int winner;

        public boolean g1;

        public double [] x=new double[m];
        public double [] p=new double[m];
```

```
public double [][] t=new double[m][nx];
public double [] c=new double[m];
public double [][] b=new double[nx][m];
public double [] d=new double[nx];
public int [] r=new int[nx];
public int [] exp=new int[nx];
public double rho;

//public double g1;

public Network(){

        //Training
        training();

        //test
        test();

        //test not predefined pattern
        testNoPattern();
 }

 private void testNoPattern() {
         int errorOnNoise=0;
         for(int trial=0;trial<1000;trial++){
                 for(int addNoise=0;addNoise<1000;addNoise++){
                         addNoise();
                 }
                 r[0]=0;
                 r[1]=0;
                 r[2]=0;
                 g1=true;
                 compute();
                 g1=false;
                 for(int y=0;y<10;y++)
                         compute();

                 if(r[0]==1||r[1]==1||r[2]==1){
                         errorOnNoise++;

                 }
          }
          System.out.println("No pattern"+(double)errorOnNoise/10.0);

 }

 public void training(){
        g1=true;
        float sumck;

        for(int i=0;i<m;i++){
                for(int j=0;j<nx;j++){
                        b[j][i]=E/(E+m-1);
                        t[i][j]=1;
                }
```

```
}



a();

r[0]=1;
r[1]=0;
r[2]=0;
compute();

sumck=0;
for(int k=0;k<m;k++){
        sumck+=x[k];
}

for(int i=0;i<m;i++){
        b[0][i]=E*x[i]/(E+1+sumck);
        t[i][0]=x[i];
}

b();

r[0]=0;
r[1]=1;
r[2]=0;
compute();

sumck=0;
for(int k=0;k<m;k++){
        sumck+=x[k];
}

for(int i=0;i<m;i++){
        b[1][i]=E*x[i]/(E+1+sumck);
        t[i][1]=x[i];
}

c();

r[0]=0;
r[1]=0;
r[2]=1;
compute();

sumck=0;
for(int k=0;k<m;k++){
        sumck+=x[k];
}

for(int i=0;i<m;i++){
        b[2][i]=E*x[i]/(E+1+sumck);
        t[i][2]=x[i];
}
```

```
        for(int i=0;i<m;i++){
                c[i]=0;
        }

        for(int j=0;j<nx;j++){
                r[j]=0;
        }

}

//Evaluation of the net
private void compute() {

        int sumc=0;
        int sumx=0;
        for(int i=0;i<m;i++){

                p[i]=0;
                for(int j=0;j<nx;j++){
                        p[i]+=t[i][j]*r[j];
                }

                if(p[i]>0.5){
                        p[i]=1;

                }else{
                        p[i]=0;
                }
                if(g1){
                        c[i]=x[i];
                }else{
                        if((p[i]+x[i])>=2.0){
                                c[i]=1;

                        }else{
                                c[i]=0;
                        }
                }
                if(c[i]==1){
                        sumc++;
                }

                if(x[i]==1){
                        sumx++;
                }
        }

        if((((double)sumc)/((double)sumx))<pho0){

                for(int j=0;j<nx;j++){
                        r[j]=0;
                }

        }else{
                double max=Double.MIN_VALUE;
```

```
                int rmax=-1;

                for(int i=0;i<nx;i++){

                        d[i]=0;
                        for(int j=0;j<m;j++){
                                d[i]+=b[i][j]*c[j];
                        }
                }

                for(int i=0;i<nx;i++){
                        if(d[i]>0.5 &&d[i]>max){
                                max=d[i];
                                rmax=i;
                        }
                }

                for(int i=0;i<nx;i++){
                        if(i==rmax){
                                r[i]=1;

                        }else{
                                r[i]=0;

                        }
                }
        }


}


//Select a char
private void selectAchar() {
        if(Math.random()<0.33)
                a();
        else
                if(Math.random()<0.5)
                        b();
                else
                        c();
}


//add a bit of noise
private void addNoise() {
        int change=(int)(Math.random()*35.99);
        x[change]=1-x[change];
}


//Test 100 input with increasing noise
public void test(){
        for(int noise=0;noise<50;noise++){
```

```
                int errorOnNoise=0;
                for(int trial=0;trial<1000;trial++){


                        selectAchar();
                        //Add noise
                        for(int addNoise=0;addNoise<noise;addNoise++){
                                addNoise();
                        }
                        r[0]=0;
                        r[1]=0;
                        r[2]=0;
                        g1=true;
                        compute();
                        g1=false;
                        for(int y=0;y<10;y++)
                                compute();

                        for(int e=0;e<nx;e++){
                                if(exp[e]!=r[e]){
                                        errorOnNoise++;
                                        break;
                                }
                        }

                }
                System.out.println(noise+","+(double)errorOnNoise/10.0);
        }
}


public void a(){

        //  **
        // *  *
        //*    *
        //******
        //*    *
        //*    *
        x[0]=0;x[1]=0;x[2]=1;x[3]=1;x[4]=0;x[5]=0;
        x[6]=0;x[7]=1;x[8]=0;x[9]=0;x[10]=1;x[11]=0;
        x[12]=1;x[13]=0;x[14]=0;x[15]=0;x[16]=0;x[17]=1;
        x[18]=1;x[19]=1;x[20]=1;x[21]=1;x[22]=1;x[23]=1;
        x[24]=1;x[25]=0;x[26]=0;x[27]=0;x[28]=0;x[29]=1;
        x[30]=1;x[31]=0;x[32]=0;x[33]=0;x[34]=0;x[35]=1;
        exp[0]=1;
        exp[1]=0;
        exp[2]=0;

}


public void b(){
```

```
            // ***
            // *  *
            // ****
            // *   *
            // *   *
            // *****

            x[0]=0;x[1]=1;x[2]=1;x[3]=1;x[4]=0;x[5]=0;
            x[6]=0;x[7]=1;x[8]=0;x[9]=0;x[10]=1;x[11]=0;
            x[12]=0;x[13]=1;x[14]=1;x[15]=1;x[16]=1;x[17]=0;
            x[18]=0;x[19]=1;x[20]=0;x[21]=0;x[22]=0;x[23]=1;
            x[24]=0;x[25]=1;x[26]=0;x[27]=0;x[28]=0;x[29]=1;
            x[30]=0;x[31]=1;x[32]=1;x[33]=1;x[34]=1;x[35]=1;
            exp[0]=0;
            exp[1]=1;
            exp[2]=0;

        }

        public void c(){
            // ****
            //*    *
            //*
            //*
            //*    *
            // ****

            x[0]=0;x[1]=1;x[2]=1;x[3]=1;x[4]=1;x[5]=0;
            x[6]=1;x[7]=0;x[8]=0;x[9]=0;x[10]=0;x[11]=1;
            x[12]=1;x[13]=0;x[14]=0;x[15]=0;x[16]=0;x[17]=0;
            x[18]=1;x[19]=0;x[20]=0;x[21]=0;x[22]=0;x[23]=0;
            x[24]=1;x[25]=0;x[26]=0;x[27]=0;x[28]=0;x[29]=1;
            x[30]=0;x[31]=1;x[32]=1;x[33]=1;x[34]=1;x[35]=0;
            exp[0]=0;
            exp[1]=0;
            exp[2]=1;


        }

}
```

9.B. ART-I Case Study: Speech Recognition†

9.B.1. *Input matrix set-up for spoken words*

The speech recognition problem considered here is one of distinguishing between three spoken words: "five", "six" and "seven". Under the present design, the above words, once spoken, are passed through an array of five band pass filters and the energy of the outputs of each of these filters is first averaged over intervals of 20 milliseconds over 5 such segments totaling 100 milliseconds. The power (energy) is compared against a weighted threshold at each frequency band to yield a 5×5 matrix of 1's and 0's that corresponds to each uttered word of the set of words considered, as shown below. The reference input matrix is obtained by repeating each of the three words 20 times and averaging the power at each frequency band per each over 20 millisecond time segments.

9.B.2. *Simulation programs Set-Up*

EXECUTION PROGRAM a:art100.exe

Text of the program written in C a:art100.cpp

To use this program:

Display

"5", "6" or "7" (zero-random noise) – choose input pattern (patterns are in three groups:

5 patterns which represents word "5" when uttered in different intonations:

"6" -similar to "5"

"7" -similar to "6"

Pattern # (0-random) – there are 10 different input patterns representing words from the set of words "5", "6" or "7", so choose one

Create new pattern for: – specify to which number the new pattern should be assigned

TERMINATION OF PROGRAM:

When the program does not ask for some input, press any key but for the space key.

When pressing space and not being asked for some input, the program will continue.

The variables used in the program are as follows:

PATT – stored patterns

PPATT – previous inputs associated with the patterns stored in the comparison layer

(used when updating old patterns)

T – weights associated with the comparison layer neurons

†Computed by Hubert Kordylewski, EECS Dept., University of illinois, Chicago, 1993.

TO – weights of a neuron in the comparison layer associated with the winning neuron form the recognition layer

TS – status of the recognition layer neurons (inhibited from firing −1, no inhibition −0)

BO – input to the recognition layer neurons (dot product between input and weights in the recognition layer)

C – outputs form the recognition layer

INP – input vector

NR – the number of patterns stored in weights of the recognition and comparison layers

GAIN – 1-when a stored pattern matches with input and 2-when input does not match with any stored pattern

SINP – number of "1"s in the input vector

SC – number of "1"s in the comparison layer output

STO – number of "1"s in the chosen pattern form (patterns are stored in the weights of the comparison layer)

MAXB – pointer to chosen pattern which best matches the input vector

The program's flow chart is given in Fig. 9.B.1.

We comment that in the ART program below the measure of similarity of ART-I and which is denoted as D, is modified from its regular ART-I form, to become

$$D\text{ (modified)} = \min(D, D1)$$

where D is the regular D of ART-I and

$$D1 = c/p; \quad p = \text{number of 1's in chosen pattern}$$

Example:

$$\begin{aligned} \text{Input vector} \quad & 1111000000; x = 4 \\ \text{Chosen pattern} \quad & 1111001111; p = 8 \\ \text{Comparison layer} \quad & 1111000000; c = 4 \end{aligned}$$

to yield:

$$D = c/x = 4/4 = 1.0 \text{ in regular ART-I}$$

$$D1 = c/p = 4/8 = 0.5$$

$$D\text{ (modified)} = \min(D, D1) = 0.5$$

This modification avoids certain difficulties in recognition in the present application.

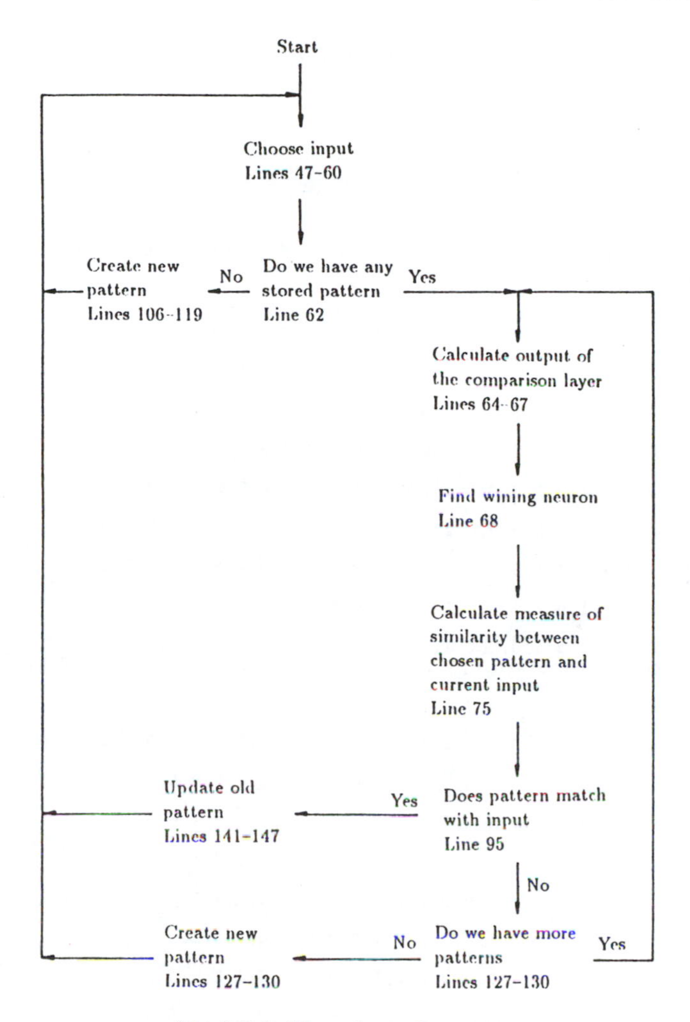

Fig. 9.B.1. Flow chart of program.

9.B.3. *Computer simulation of ART program (C-language)*

```
#include <stdio.h>
#include <stdlib.h>
#include <math.h>
#include <conio.h>
main()
{
char ch;
int l,k,j,np=0,i,t[8][5][5],to[5][5],ts[8],c[5][5],tm[5][5],inp[5][5];
int gain2=2,sinp=0,sc=0,maxb=0;
int maxi=0,sto,ss,x,y,kk=0,fp,ks[10];
int patt7[10][5][5]={0,0,0,1,0,1,0,0,0,1,0,1,0,0,0,0,0,0,1,1,1,0,0,0,0,
                     0,0,0,1,0,1,0,0,0,1,0,0,0,0,0,0,0,0,1,1,1,0,0,0,0,

         0,0,0,1,0,1,0,0,0,1,0,0,0,0,0,0,0,0,1,1,1,0,0,0,0,

         0,0,0,1,1,0,0,0,0,1,0,0,0,0,0,0,0,0,1,0,0,0,0,0,0,

         0,0,0,1,0,1,0,0,0,1,0,1,0,0,0,0,0,0,0,1,1,0,0,0,0,

         0,0,0,1,0,1,0,0,0,1,0,1,0,0,0,0,0,0,0,1,1,0,0,0,0,

         0,0,0,1,0,1,0,0,0,1,0,1,0,0,0,0,0,0,1,0,1,0,0,0,0,

         0,0,0,1,0,1,0,0,0,1,0,0,0,0,0,0,0,0,1,1,1,0,0,0,0,

         0,0,0,1,0,1,0,0,0,1,0,0,0,0,0,0,0,0,1,1,1,0,0,0,0,

         0,0,0,1,0,1,0,0,0,1,0,1,0,0,0,0,0,0,1,1,1,0,0,0,0};
int patt6[10][5][5]={0,0,0,1,0,0,0,1,0,0,0,1,1,0,0,0,1,0,0,0,0,0,0,0,0,

         0,0,0,1,0,0,0,1,0,0,0,1,1,0,0,0,1,0,0,0,1,0,0,0,0,

         0,0,0,1,0,0,0,1,0,0,0,1,1,0,0,0,1,0,1,1,1,0,0,0,0,

         0,0,0,1,0,0,0,1,0,0,0,1,1,0,0,0,1,0,0,0,0,0,0,0,0,

         0,0,0,1,0,0,0,1,0,0,0,1,1,0,0,0,1,0,0,0,1,0,0,0,1,

         0,0,0,1,0,0,0,1,0,0,0,1,1,0,0,0,1,0,1,0,1,0,0,0,1,

         0,0,0,1,0,0,0,1,0,0,0,1,1,0,0,0,0,0,0,1,1,0,0,0,0,

         0,0,0,1,0,0,0,1,0,0,0,1,1,0,0,0,1,0,0,0,1,0,0,0,0,

         0,0,0,1,0,0,0,1,0,0,0,1,1,0,0,0,1,0,1,1,1,0,0,0,1,

         0,0,0,1,0,0,0,1,0,0,0,1,1,0,0,0,1,0,0,1,1,0,0,0,1};
int patt5[10][5][5]={0,1,0,0,0,1,0,0,0,1,1,0,0,1,0,0,0,1,0,0,1,0,0,0,0,

         0,1,0,0,0,1,0,0,0,1,1,0,0,0,0,0,0,1,1,0,0,0,0,0,0,

         0,0,1,0,0,1,0,0,0,1,1,0,0,0,0,0,0,1,0,0,0,0,0,0,0,

         0,0,1,0,0,1,0,0,0,1,1,0,0,0,0,0,0,1,0,0,0,0,0,0,0,

         0,1,0,0,0,1,0,0,0,1,0,0,0,1,0,0,1,1,0,0,0,1,0,0,0,

         0,0,1,0,0,1,0,0,0,1,1,0,0,0,0,0,0,0,0,0,0,0,0,0,0,

         0,1,0,0,0,1,0,0,0,1,1,0,0,1,0,0,0,1,0,0,0,0,0,0,0,

         0,1,0,0,0,1,0,0,0,1,1,0,0,0,0,0,0,1,0,0,0,0,0,0,0,

         0,1,0,0,0,1,0,0,0,1,1,0,0,0,0,0,0,1,0,0,0,0,0,0,0,

         0,1,0,0,0,1,0,0,0,1,1,0,0,1,0,0,0,1,0,0,0,0,0,0,0};
float b[8][5][5],bo[8],bt,ro,rol,rp;
int nr,patt[10][5][5],ppatt[10][5][5];
clrscr();
for(i=0;i<8;i++){for(j=0;j<5;j++){for(k=0;k<5;k++){
```

```
t[i][j][k]=1;b[i][j][k]=0.077;ppatt[i][j][k]=1;}}}
do{clrscr();maxb=0;
        for (i=0;i<8;i++){ts[i]=0;bo[i]=0;}
        for (i=0;i<5;i++){for (j=0;j<5;j++){to[i][j]=0;c[i][j]=0;}}
        for(i=0;i<5;i++){for(j=0;j<5;j++){inp[i][j]=0;}}
        printf("5 or 6 or 7 (0-random #):");scanf("%d", &nr);
        if(nr==0){nr=rand()%3+5;printf("nr=%d\n", nr);}
        printf("Pattern #(0-random #):");scanf("%d", &kk);
        if(kk==0) {kk=rand()%10+1;printf("kk=%d\n", kk);ch=getche();}
        if(nr<6){for(i=0;i<5;i++){for(j=0;j<5;j++)
                                                                         patt[kk-
1][i][j]=patt5[kk-1][i][j];}}
        if(nr>6){for(i=0;i<5;i++){for(j=0;j<5;j++)
                                                                         patt[kk-
1][i][j]=patt7[kk-1][i][j];}}
        if(nr==6){for(i=0;i<5;i++){for(j=0;j<5;j++)
                                                                         patt[kk-
1][i][j]=patt6[kk-1][i][j];}}
        if(kk<11){fp=patt[kk-1][0][0];
        for(i=0;i<5;i++){for(j=0;j<5;j++){if((patt[kk-1][i][j]+fp)>0)

                 inp[i][j]=1;

                 fp=patt[kk-1][i][j]; }}
                                 }
               else   do{
        printf("x:");scanf("%d", &x);printf("y:");scanf("%d", &y);
        inp[x][y]=1;printf("next press A\n");printf("\n");
        ch=getche();
        }while(ch=='A');
        do{clrscr();if (np>0)
                          {
                          gain2=2;
                          for (i=0;i<5;i++) ts[i]=0;
                          for (i=0;i<np;i++){
                          for (j=0;j<5;j++){for(k=0;k<5;k++){
c[j][k]=inp[j][k];to[j][k]=0;}}
                          for (j=0;j<5;j++) bo[j]=0;
                          for (j=0;j<np;j++) {  for (k=0;k<5;k++){
                          for(x=0;x<5;x++){bo[j]=bo[j]+c[k][x]*b[j][k][x];}}}
                 for (j=0;j<np;j++) {if(ts[j]<1) {maxb=j;j=5;}}
                 for (j=0;j<np;j++) {if(ts[j]<1){if (bo[maxb]<bo[j])
                                              maxb=j;}}
          for (j=0;j<5;j++) {printf("bo:%f   ts:%d\n", bo[j], ts[j]);}
          for (j=0;j<5;j++){for(k=0;k<5;k++)to[j][k]=t[maxb][j][k];}
                                  printf("\n");
          for (j=0;j<5;j++){for(k=0;k<5;k++)
                 {if(inp[j][k]+to[j][k]>1)c[j][k]=1; else c[j][k]=0;}}
          sc=0;sinp=0;sto=0;
          for (j=0;j<5;j++){for(k=0;k<5;k++)
                 {sto=sto+to[j][k];sc=sc+c[j][k]; sinp=sinp+inp[j][k];}}
          printf("Stored patterns \n");
          for(j=0;j<5;j++){for(k=0;k<np;k++){
printf("%d%d%d%d%d    ",t[k][j][0],t[k][j][1],t[k][j][2],t[k][j][3],t[k][j][4]);}
                                  printf("\n");}
          printf("Max bo has pattern nr:%d\n\n", maxb+1);
          printf("Pattern        Input          C\n");printf(" ");
          for(j=0;j<5;j++){
printf("%d%d%d%d%d        ",to[j][0],to[j][1],to[j][2],to[j][3],to[j][4]);
printf("%d%d%d%d%d        ",inp[j][0],inp[j][1],inp[j][2],inp[j][3],inp[j][4]);
printf("%d%d%d%d%d       \n ",c[j][0],c[j][1],c[j][2],c[j][3],c[j][4]);}
       /*  if((abs(sc-sinp)>1)||(abs(sto-sinp)>1)||(abs(sto-sc)>1))*/
                          ro1=sinp;
                          ro=sc/ro1;
                          ro1=sto;rp=sc/ro1;
printf("# of '1' C:%d  Pattern (T):%d Input (I):%d C/I:%f C/P:%f\n", sc, sto,
sinp, ro, rp);
if(rp<ro)ro=rp;
                          if(ro<0.62)
                          {gain2=2;ss=0; ts[maxb]=1;printf(" \n");
```

```
                        printf("INPUT  DOES NOT MATCH WITH PATTERN #%d\n",
maxb+1);
                         for(j=0;j<5;j++)ss=ss+ts[j];if(ss==np)i=np;
                         ch=getche();clrscr();}
                        else{ gain2=1;printf(" \n");i=np;
                        printf("INPUT MATCHES WITH PATTERN #%d\n", maxb+1);
                        ch=getche();clrscr();}
                                                                }
                 }       ;clrscr();
        }while (ch=='A');
        if (np==0){printf("\n");printf("Create new pattern for #:");
                         scanf("%d", &ks[np]);
                         sc=0;for(j=0;j<5;j++){for(k=0;k<5;k++)
                         {c[j][k]=inp[j][k];sc=sc+c[j][k];}}np=np+1;
                 printf("\n");
                                        printf("Create new pattern :\n");
         for (i=0;i<5;i++){
printf("%d%d%d%d%d\n",inp[i][0],inp[i][1],inp[i][2],inp[i][3],inp[i][4]);
                 for(j=0;j<5;j++){
                t[0][i][j]=inp[i][j];ppatt[0][i][j]=inp[i][j];
                 bt=(1+sc);b[0][i][j]=2*c[i][j]/bt;}}
                 gain2=2;  } else{if(gain2==2) {
                 printf("Create new pattern for #:");scanf("%d", &ks[np]);

                                                  for(i=0;i<5;i++){
printf("%d%d%d%d%d\n",inp[i][0], inp[i][1], inp[i][2],inp[i][3],inp[i][4]);
                                                  for(j=0;j<5;j++){
                        -        t[np][i][j]=inp[i][j];
                                 ppatt[np][i][j]=inp[i][j];
                          bt=(1+sinp);b[np][i][j]=2*inp[i][j]/bt;}}
                                   np=np+1;}
                                          printf("\n");
                        if(gain2==1) {printf("Previous pattern:       Last
inp:\n");
                 sto=0;   for (i=0;i<5;i++){
printf("%d%d%d%d%d
",to[i][0],to[i][1],to[i][2],to[i][3],to[i][4]);
printf("    %d%d%d%d%d\n",ppatt[maxb][i][0],ppatt[maxb][i][1],ppatt[maxb][i][2],
ppatt[maxb][i][3],ppatt[maxb][i][4]);
                          for (j=0;j<5;j++){
                          if(t[maxb][i][j]+inp[i][j]>0)tm[i][j]=1;
                          else tm[i][j]=0;
                          if(ppatt[maxb][i][j]+tm[i][j]>1)
                          {t[maxb][i][j]=1;sto=sto+1;}
                          else t[maxb][i][j]=0;

                          /*
                    if(ppatt[maxb][i][j]==0)   {t[maxb][i][j]=c[i][j];
                  bt=(1+sc);b[maxb][i][j]=2*c[i][j]/bt;}
                if(ppatt[maxb][i][j]==1)  { t[maxb][i][j]=inp[i][j];
                     bt=(1+sinp);b[maxb][i][j]=2*inp[i][j]/bt;}*/
                    ppatt[maxb][i][j]=inp[i][j];c[i][j]=t[maxb][i][j];}
                                 }
                         printf("\n");
                printf("Updated  pattern:       Input:        \n");
                for(j=0;j<5;j++)              {
printf("%d%d%d%d%d                 ",c[j][0],c[j][1],c[j][2],c[j][3],c[j][4]);
printf("%d%d%d%d%d   \n",inp[j][0],inp[j][1],inp[j][2],inp[j][3],inp[j][4]);
                 }printf("\n");
      printf("INPUT:%d(%d)                FOUND:%d\n", nr, kk, ks[maxb]);
                 } }
                 for(i=0;i<5;i++){for(j=0;j<5;j++){bt=(1+sto);
                 b[maxb][i][j]=2*t[maxb][i][j]/bt;}
                 }
        ch=getche();
}while (ch==' ');
}
```

9.B.4. *Simulation results*

```
(1)    Input:    01100
                 11001  - ASSUMED INPUTS
                 10011
                 01110
                 01100

       bo:1.571429   ts:0
       bo:0.600000   ts:0
       bo:0.923077   ts:0
       bo:1.400000   ts:0
       bo:0.000000   ts:0

       Stored patterns
       01100    00011    00011    00110
       11001    00110    11001    11001
       11011    01110    11100    11000
       00110    01100    00011    00110    namely: ('five', 'six', 'seven',
       11000    00000    11000    00000    'none of the above')
       Max bo has pattern nr:1

       Pattern             Input         C = Pattern Input
        01100              01100        01100
        11001              11001        11001
        11011              10011        10011
        00110              01110        00110
        11000              01100        01000
        # of '1' C:11 Pattern (T):13 Input (I):13 C/I:0.8461 C/P:0.846154

       INPUT MATCHES WITH PATTERN #1

       Previous pattern:         Last inp:
       01100                     01100
       11001                     11001
       11011                     11011
       00110                     00110
       11000                     11000

       Updated pattern:          Input:
       01100                     01100
       11001                     11001
       1①011                     1⓪011             encircled points: where
       0⓪110                     0⓪110             updated pattern
       ①1000                     ⓪1①00             differs from input

       INPUT:5(5)                FOUND:5
```

```
(2)    Input:    01100
                 11001
                 10011
                 01110
                 01100

       bo:1.571429   ts:0
       bo:0.600000   ts:0
       bo:0.923077   ts:0
       bo:1.400000   ts:0
       bo:0.000000   ts:0

       Stored patterns
       01100    00011    00011    00110
       11001    00110    11001    11001
       11011    01110    11100    11000
       00110    01100    00011    00110
       11000    00000    11000    00000
       Max bo has pattern nr:1

       Pattern            Input        C
        01100             01100        01100
        11001             11001        11001
        11011             10011        10011
        00110             01110        00110
        11000             01100        01000
        # of '1' C:11 Pattern (T):13 Input (I):13 C/I:0.846 C/P:0.846154

       INPUT MATCHES WITH PATTERN #1

       Previous pattern:       Last inp:
       01100                   01100
       11001                   11001
       11011                   10011
       00110                   01110
       11000                   01100

       Updated pattern:        Input:
       01100                   01100
       11001                   11001
       10011                   10011
       01110                   01110
       01100                   01100

       INPUT:5(5)              FOUND:5
```

These are identical since input was the same.
Updated pattern = 1 only if: LAST INP = 1 and [(INPUT) or (PATTERN)] = 1.

Chapter 10

The Cognitron and the Neocognitron

10.1. Background of the Cognitron

The cognitron, as its name implies, is a network designed mainly with recognition of patterns in mind. To do this, the cognitron network employs both *inhibitory* and *excitory* neurons in its various layers. It was first devised by Fukushima (1975), and is an *unsupervised* network such that it resembles the biological neural network in that respect.

10.2. The Basic Principles of the Cognitron

The cognitron basically consists of layers of inhibitory and excitory neurons. Interconnection of a neuron in a given layer is only to neurons of the previous layer that are in the vicinity of that neuron. This vicinity is termed as the *connection competition region* of the given neuron. For training efficiency, not all neurons are being trained. Training is thus limited to only an *elite group* of the most relevant neurons, namely to neurons already previously trained for a related task.

Whereas connection regions lead to *overlaps* of neurons, where a given neuron may belong to the connection region of more than one upstream neuron, *competition* (for "elite" selection) is introduced to overcome the effect of the overlaps. Competition will disconnect the neurons whose responses are weaker. The above feature provides the network with considerable redundancy, to enable it to function well in the face of "lost" neurons.

The cognitron's structure is based on a multi-layer architecture with a progressive reduction in number of competition regions. Alternatively, groups of two layers, L-I and L-II may be repeated n times to result in $2n$ layers in total (L-I_1, L-II_1, L-I_2, L-II_2, etc.).

10.3. Network Operation

(a) Excitory Neurons
The output of an *excitory neuron* is computed as follows:

Let y_k be the output from an excitory neuron at the previous layer and let v_j be the output from an inhibitory neuron at its previous layer. Define the output components of the excitory ith neuron as:

$$x_i = \sum_k a_{ik} y_k \qquad \text{due to excitation inputs} \tag{10.1}$$

$$z_i = \sum_k b_{ik} v_k \qquad \text{due to inhibition inputs} \tag{10.2}$$

a_{ik} and b_{ik} being relevant weights, that are adjusted when the neuron concerned is more active than its neighbors, as discussed in 10.4 below. The total output of above neuron is given as:

$$y_i = f(N_i) \tag{10.3}$$

where

$$N_i = \frac{1 + x_i}{1 + z_i} - 1 = \frac{x_i - z_i}{1 + z_i} \tag{10.4}$$

$$f(N_i) = \begin{cases} N_i \cdots & \text{for} \quad N_i \geq 0 \\ 0 \cdots & \text{for} \quad N_i < 0 \end{cases} \tag{10.5}$$

Hence, for small z_i

$$N_i \cong x_i - z_i \tag{10.6}$$

However, for very large x, z

$$N_i = \frac{x_i}{z_i} - 1 \tag{10.7}$$

Furthermore, if both x and z increase linearly with some γ namely:

$$x = p\gamma \tag{10.8}$$

$$z = q\gamma \tag{10.9}$$

p and q being constants, then

$$y = \frac{p - q}{2q} \left[1 + \tanh\left(\log \frac{pq}{2}\right)\right] \tag{10.10}$$

which is of the form of the Weber–Fechner law (See: Guyton, 1971, pp. 562–563) that approximates responses of biological sensory neurons.

(b) Inhibitory Neurons

The output of an inhibitory neuron is given by:

$$v = \sum_i c_i y_i \tag{10.11}$$

where

$$\sum_i c_i = 1 \tag{10.12}$$

y_i being an output of an excitory cell. The weights c_i are *preselected* and do not undergo modification during network training.

10.4. Cognitron's Network Training

The a_{ji} weights of the excitory neuron in a two-layer cognitron structure are iterated by δa as in Eq. (10.13) but *only* if that neuron is a winning neuron in a region, where a_{ji} is as in Eq. (10.1) (namely, a_{ji} is the weight on an excitory input y_j to the given excitory neuron), and c_j being the weight at the input to the inhibitory neuron of this layer, whereas q is a preadjusted learning (training) rate coefficient (see Fig. 10.1).

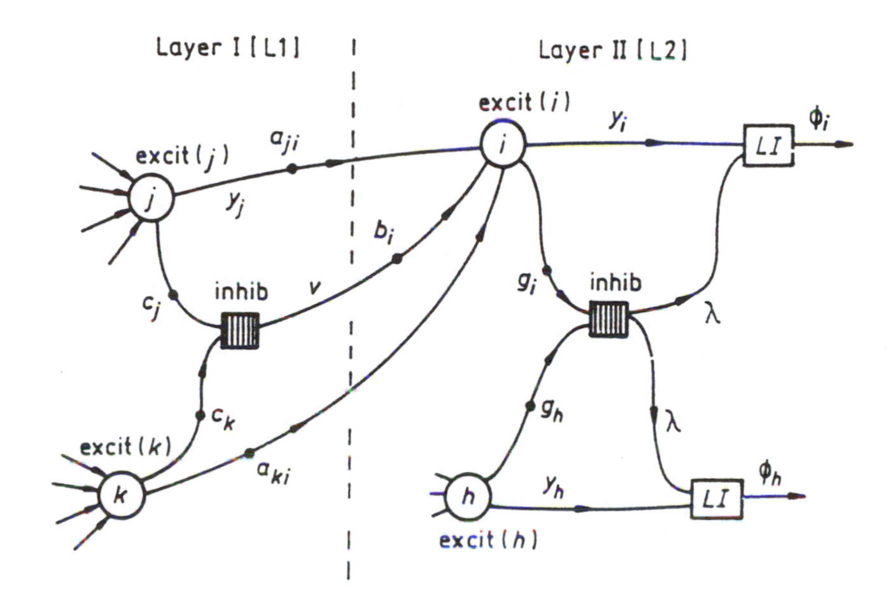

Fig. 10.1. Schematic description of a cognitron network (a competition region with two excitory neurons in each layer).

$$\partial\, a_{ji} = q\; c_j^*\; y_j^* \quad \text{(asterisk denoting previous layer)} \tag{10.13}$$

Note that there are several *excitory* neurons in each competition region of layer L1 and only one inhibitory layer.

The inhibitory weights b_j to excitory neurons are iterated according to:

$$\partial b_i = \frac{q \sum_j a_{ji} y_j^z}{2v^*}\,; \quad \partial b_i = \text{change in } b_i \tag{10.14}$$

where b_i are the weights on the connection between the inhibitory neuron of layer L1 and the ith excitory neuron in L2, Σ_j denoting the summation on weights from all excitory L1 neurons to the same ith neurons of L2, while v is the value of the inhibitory output as in Eq. (10.11), q being a rate coefficient.

If *no neuron* is active in a given competition region, then Eqs. (10.13), (10.14) are replaced by (10.15), (10.16), respectively:

$$\partial a_{ji} = q' c_j y_j \tag{10.15}$$

$$\partial b_i = q' v_i \tag{10.16}$$

where

$$q' < q \tag{10.17}$$

such that now the higher the inhibition output, the higher is its weight, in sharp contrast to the situation according to Eq. (10.13).

Initialization

Note that initially all weights are 0 and no neuron is active (providing an output). Now, the first output goes through since at the first layer of excitory neurons the network's input vector serves as the y vector of inputs to L1, to start the process via Eq. (10.15) above.

Lateral Inhibition

An inhibitory neuron is also located in each competition region as in layer L2 of Fig. 10.1 to provide lateral inhibition whose purpose (not execution) is as in the ART network of Chap. 9 above. This inhibitory neuron receives inputs from the excitory neurons of its layer via weights g_i. It's output λ is given:

$$\lambda = \sum_i g_i y_i \tag{10.18}$$

y_i being the outputs of the excitory neuron of the previous (say, L1) layer, and

$$\sum_i g_i = 1 \tag{10.19}$$

Subsequently the output λ of the L2 inhibitory neuron above modifies the actual output of the ith L2 excitory neuron from y_i to ϕ_i where

$$\phi_i = f\left[\frac{1 + y_i}{1 + \lambda} - 1\right] \tag{10.20}$$

where y_i are as in Eqs. (10.3) and (10.5) above, $f(\ldots)$ being as in Eq. (10.5), resulting in a feedforward form of lateral inhibition and which is applicable to all layers.

10.5. The Neocognitron

A more advanced version of the cognitron also developed by Fukushima *et al.* (1983), is the neocognitron. It is hierarchical in nature and is geared toward simulating human vision. Specific algorithms for the neocognitron are few and very complex, and will therefore not be covered in this text.

The recognition is arranged in a hierarchical structure of groups of 2 layers, as in the case of the cognitron. The two layers now are a (simple-cells-) layer (S-layer) and a concentrating layer (C-layer), starting with an S-layer denoted as S1 and ending with a C layer (say, C4). Each neuron of the S-layer responds to a given feature of its input layers (including the overall network's input). Each of the arrays of the C layer processes in depth inputs from usually one S layer array.

The number of neurons and arrays generally goes down from layer to layer. This structure enables the neocognitron to overcome recognition problems where the original cognitron failed, such as images under position or angular distortions (say somewhat rotated characters or digits in handwriting recognition problems).

See Fig. 10.2.

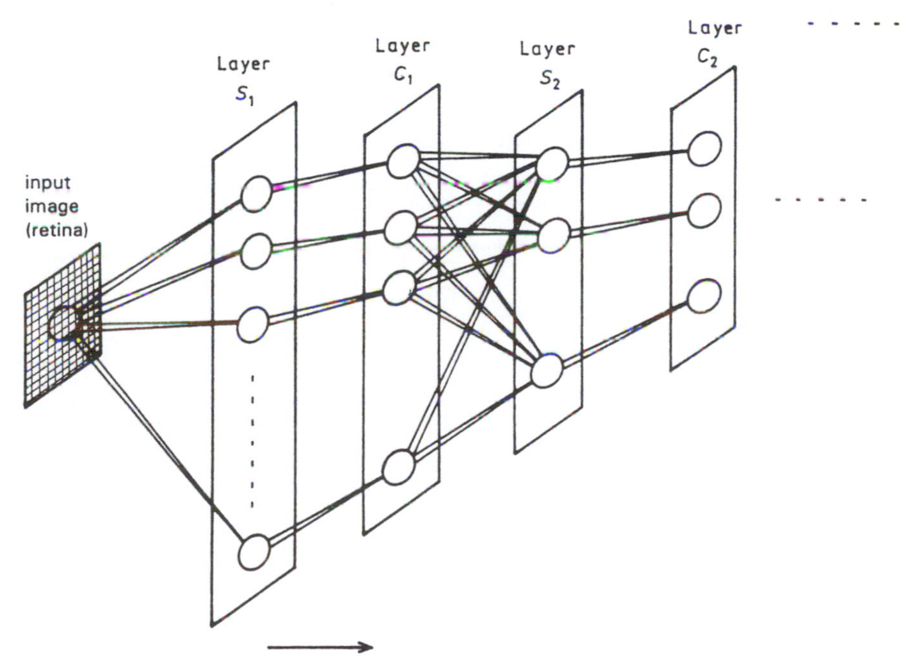

Fig. 10.2. Schematic of a neocognitron.

Chapter 11

Statistical Training

11.1. Fundamental Philosophy

The fundamental idea behind statistical (stochastic) training of neural networks is: Change weights by a random small amount and keep those changes that improve performance.

The weakness of this approach is that it is extremely *slow*! Also, it can get stuck at a *local minimum* if random changes are small since the change may not have enough power to climb "over a hill" (see Fig. 11.1) in order to look for another valley.

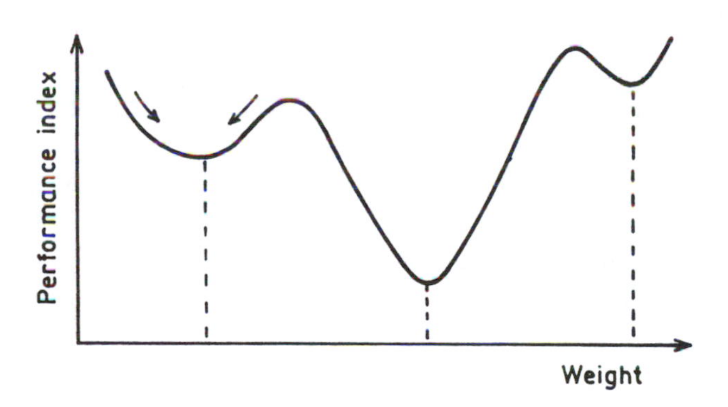

Fig. 11.1. A performance cost with many minima.

To overcome getting stuck in a local minimum, *large weight changes* can be used. However, then the network may become oscillatory and miss settling at any minimum. To avoid this possible *instability*, weight changes can be *gradually decreased in size*. This strategy resembles the processes of annealing in metallurgy. It basically applies to all networks described earlier, but in particular to back propagation and modified networks.

11.2. Annealing Methods

In metallurgy, annealing serves to obtain a desired mixing of molecules for forming a metal alloy. Hence, the metal is initially raised to a temperature above its melting point. In that liquid state the molecules are shaken around wildly, resulting in a high distance of travel. Gradually the temperature is reduced and consequently the amplitude of motion is reduced until the metal settles at the lowest energy level. The motion of molecules is governed by a Boltzman probability distribution.

$$p(e) = \exp(-e/KT) \tag{11.1}$$

Where $p(e)$ is the probability of the system being at energy level e. K being the Boltzman constant, T denoting absolute temperature in Kelvin degrees (always positive). In that case, when T is high, $\exp(-e/KT)$ approaches zero, such that almost any value of e is probable, namely is $p(e)$ is high for any relatively high e. However, when T is reduced, the probability of high values of e is reduced since e/KT increases such that $\exp(-e/KT)$ is reduced for high e.

11.3. Simulated Annealing by Boltzman Training of Weights

We substitute for e of Eq. (11.1) with ΔE which denotes a change in the energy function E

$$p(\Delta E) = \exp(-\Delta E/KT) \tag{11.2}$$

while T denotes some temperature equivalent. A neural network weight training procedure will thus become:

(1) Set the temperature equivalent T at some high initial value.
(2) Apply a set of training inputs to the network and calculate the network's outputs, and compute the energy function.
(3) Apply a random weight change Δw and recalculate the corresponding output and the energy function (say a squared error function $E = \Sigma_i\ (\text{error})^2$).
(4) If the energy of the network is reduced (to indicate improved performance) then keep Δw, else: calculate the probability of $p(\Delta E)$ of accepting Δw, via Eq. (11.2) above and select some pseudo random number r from a uniform distribution between 0 and 1. Now, if $p(\Delta E) > r$ (note: $\Delta E > 0$ in the case of increase in E) then still accept the above change, else, go back to the previous value of w.
(5) Go to Step (3) and repeat for all weights of the network, while gradually reducing T after each complete set of weights has been (re-)adjusted.

The above procedure allows the system to occasionally accept a weight change in the wrong direction (worsening performance) to help avoiding it from getting stuck at the local minimum.

The gradual reduction of the temperature equivalent T may be deterministic (following a pre-determined rate as a function of the number of the iteration). The stochastic adjustment of Δw may be as in Sec. 11.4.

11.4. Stochastic Determination of Magnitude of Weight Change

A stochastic adjustment of Δw (step 3 in Sec. 11.3 above) can also follow a thermodynamic equivalent, where Δw may be considered to obey a Gaussian distribution as in Eq. (11.4):

$$p(\Delta w) = \exp\left[-\frac{(\Delta w)^2}{T^2}\right] \tag{11.3}$$

$p(\Delta w)$ denoting the probability of a weight change Δw. Alternatively $p(\Delta w)$ may obey a Boltzman distribution similar to that for ΔE. In these cases, Step 3 is modified to select the step change Δw as follows [Metropolis *et al.*, 1953].

(3.a) Pre-compute the cumulative distribution $1P(w)$, via numerical integration

$$P(w) = \int_0^w p(\Delta w) d\Delta w \tag{11.4}$$

and store $P(w)$ versus w.

(3.b) Select a random number μ from a uniform distribution over an interval from 0 to 1. Use this value of μ so that $P(w)$ will satisfy, for some w:

$$\mu = P(w) \tag{11.5}$$

and look up the corresponding w to $P(w)$ according to (11.6). Denote the resultant w as the present w_k for the given neural branch. Hence, derive

$$\Delta w_k = w_k - w_{k-1} \tag{11.6}$$

w_{k-1} being the previous weight value at the considered branch in the network.

11.5. Temperature-Equivalent Setting

We have stated that a gradual temperature reduction is fundamental to the simulated annealing process. It has been proven [Geman and Geman 1984] that for convergence to a global minimum, the rate of temperature-equivalent reduction must satisfy

$$T(k) = \frac{T_o}{\log(1+k)}; \quad k = 0, 1, 2, \ldots . \tag{11.7}$$

k denoting the iteration step.

11.6. Cauchy Training of Neural Network

Since the Boltzman training of a neural network as in Secs. 11.2 to 11.4 is very slow, a faster stochastic method based on Cauchy probability distributions was proposed by Szu (1986). The Cauchy distribution of the energy changes is given by

$$p(\Delta E) = \frac{a\,T}{T^2 + (\Delta E)^2}; \quad a = \text{ constant} \tag{11.8}$$

to result in a distribution function of longer (slower receding) tails than in the case of the Boltzman or the Gaussian distribution. Observe that for the Cauchy distribution:

$$\mathrm{var}(\Delta E) = \infty!!$$

When the Cauchy distribution is used for Δw, the resultant Δw will satisfy

$$\Delta w = \rho T \cdot \tan\left[p(\Delta w)\right] \tag{11.9}$$

ρ being a *learning rate* coefficient. Step (3) and Step (4) of the framing procedure of Sec. 11.3 will thus become:

(3.a) Select a random number n from a uniform distribution between 0 and 1 and let

$$p(\Delta w) = n \tag{11.10}$$

where $p(\Delta w)$ is in the form of Eq. (11.8) above

(3.b) Subsequently, determine Δw via Eq. (11.9) to satisfy

$$\Delta w = \rho T \cdot \tan(n) \tag{11.11}$$

where T is updated by: $T = \frac{T_o}{1+k}$ for $k = 1, 2, 3, \ldots$ in contrast to the inverse log rate of Sec. 11.5.

Note that the new algorithm for T is reminiscent of the Dvoretzky condition for convergence in stochastic approximation [Graupe, 1989].

(4) Employ a Cauchy or a Boltzman distribution in (4) of Sec. 11.3.

The above training method is faster than the Boltzman training. However, it is still very slow. Furthermore, it may result in steps in the wrong direction to cause instability. Since the Cauchy-machine may yield very large Δw, the network can get stuck. To avoid this, *hard limits* may be set. Alternatively, Δw may be squashed using an algorithm similar to that used for the *activation function*, namely:

$$\Delta w(\text{modified}) = -M + \frac{2M}{1 + \exp(-\Delta w/M)} \tag{11.12}$$

M being the hard limit on the amplitude of Δw.

11.A. Statistical Training Case Study — A Stochastic Hopfield Network for Character Recognition*

11.A.1. *Motivation*

The case study of Sec. 11.A was concerned with situations where no local minima were encountered and thus there appeared to be no benefit in a stochastic network. We now present a problem where in certain situations a stochastic network can improve on a deterministic one, since local minima do exist. Still, not always does the stochastic algorithm improve on the deterministic one even in the present case study, as is indicated by the results below.

11.A.2. *Problem statement*

The problem of the present case study is that of recognizing noisy characters. Specifically, we are attempting to identify the characters: "H", "5", "1" all presented in an 8×8 matrix. A Hopfield network is employed in the present study whose schematic is given in Fig. 11.A.1. The study compares recognition performance

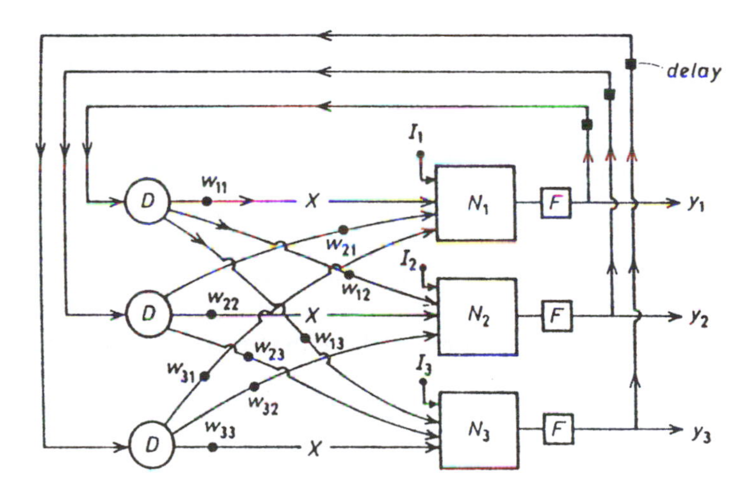

Fig. 11.A.1. The Hopfield memory.

of a deterministic Hopfield network, generally similar to that of the Case Study of Sec. 7.A, with its stochastic Hopfield network equivalent, which was simulated annealing via a Cauchy approach as in the Case Study of Sec. 11.B and via a Boltzman approach that was further discussed in Sec. 11 above.

*Computed by Sanjeev Shah, EECS Dept., University of Illinois, Chicago, 1993.

11.A.3. *Algorithm set-up*

We consider an input 8×8 matrix of elements x_i and their net values (calculated via a sigmoid function of x_i) denoted as net_i. We then follow the procedure of [cf. Freeman and Skapura 1991]:

1. Apply the incomplete or garbled vector $\bar{x}$' to the inputs of the Hopfield net.
2. Select at random an input and compute its corresponding net_k.
3. Assign $x_k = 1$ with probability of $p_k = \frac{1}{1+e^{-net_k/T}}$. Compare p_k to a number, z, taken from a uniform distribution between zero and one. If $z \leq p_k$, keep x_k.
4. Repeat 2 and 3 until all units have been selected for update.
5. Repeat 4 until thermal equilibrium has been reached at the given T. At the thermal equilibrium, the output of the units remains the same (or within a small tolerance between any two processing cycles).
6. Lower T and repeat Steps 2 to 6.

Temperatures are reduced according to Boltzman and according to Cauchy schedules. By performing annealing during pattern recall, we attempt to avoid shallow, local minima.

11.A.4. *Computed results*

The network discussed above considered inputs in an 8×8 matrix format as in Fig. 11.A.2. Recognition results are summarized in Table 11.A.1 below, as follows:

The results of Table 11.A.1 indicate that the Boltzman annealing outperformed the Cauchy annealing in all but the case of $M = 5$ exemplars. It appears that with increasing M the Cauchy annealing may improve, as also the theory seems to indicate (see Sec. 11). Also, the deterministic network outperformed the stochastic (Boltzman) network in most situations. However, in the noisy cases of $M = 4$ exemplars, the stochastic network was usually quite better. This may indicate that the deterministic network was stuck in a local minimum which the stochastic network avoided.

The Hopfield network is limited in the number of exemplars by the capacity of the Hopfield network to store pattern exemplars. In the network used, which was of 64 nodes, we could not store more than 5 exemplars, which is below the empirical low of

$$M < 0.15N \tag{11.A.1}$$

for which error in recall should be low. According to relation (11.B.1), we might have been able to reach $M = 9$.

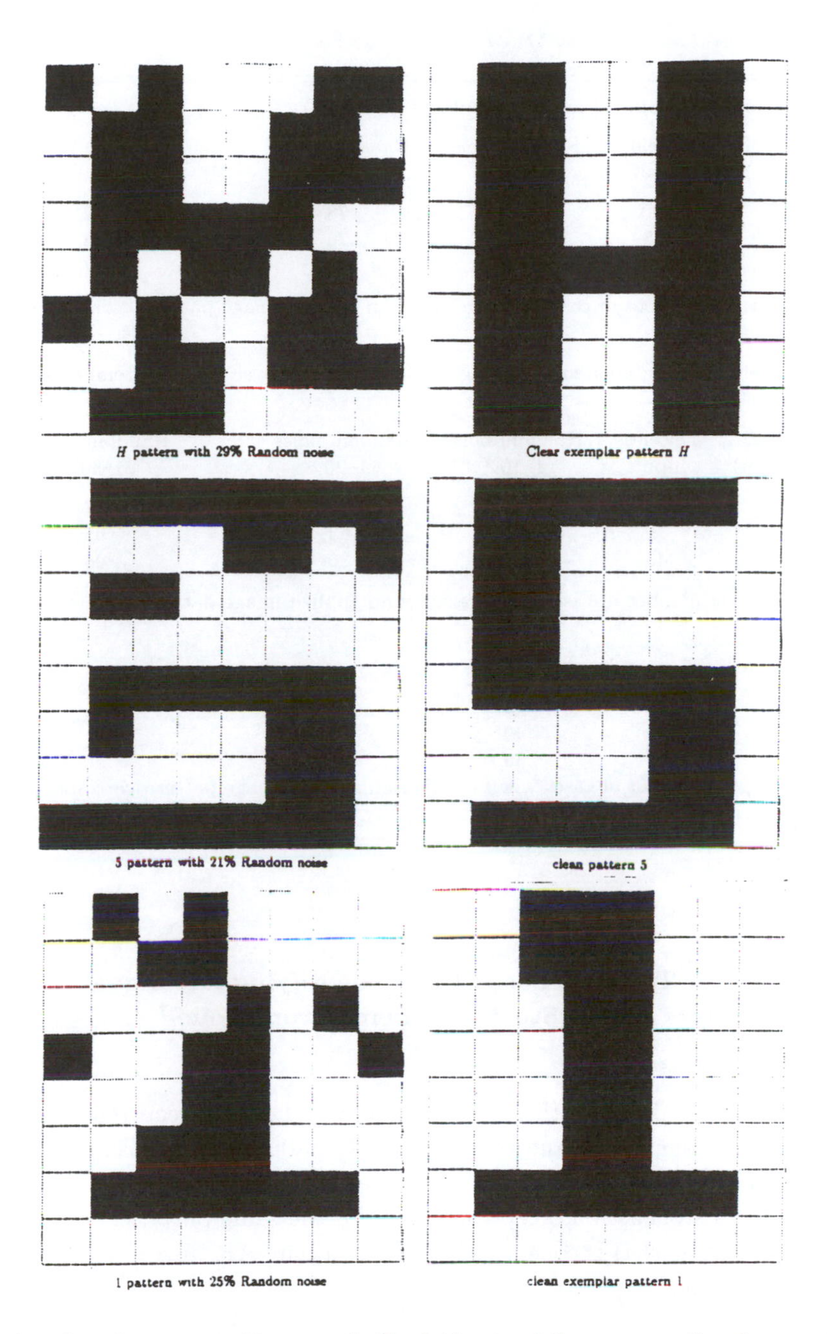

Fig. 11.A.2. Sample patterns of input to the Hopfield net and the corresponding clean versions as output by the net.

Table 11.A.1. Recognition Performance.

(a) Deterministic Hopfield net (no simulated annealing)

Number of exemplars(m)	Noiseless input	Random noise 1–10%	Random noise 11–20%	Random noise 21–30%
3	100	67	78	67
4	75	66	76	54
5	87	80	45	47

In percent, the number of cases when recall of an applied image pattern corresponded to its exemplar

(b) Hopfield net — simulated annealing — Boltzman approach Starting temperature 6, iterations 150

Number of exemplars(m)	Noiseless input	Random noise 1–10%	Random noise 11–20%	Random noise 21–30%
3	44	26	34	17
4	75	100	63	95
5	43	35	36	28

In percent, the number of cases when recall of an applied image pattern corresponded to its exemplar

(c) Hopfield net — simulated annealing — Cauchy approach

3	33	5	14	1
4	38	49	7	25
5	37	35	36	29

In percent, the number of cases when recall of an applied image pattern corresponded to its exemplar.

11.B. Statistical Training Case Study: Identifying AR Signal Parameters with a Stochastic Perceptron Model†

11.B.1. *Problem set-up*

This case study parallels the case study of Sec. 4.A of a recurrent Perceptron-based model for identifying autoregressive (AR) parameters of a signal. Whereas in Sec. 4.A a deterministic Perceptron was used for identification, the present case study employs a stochastic Perceptron model for the same purpose.

Again the signal $x(n)$ is considered to satisfy a pure AR time series model given by the AR equation:

$$x(n) = \sum_{i=1}^{m} a_i x(n-i) + w(n) \tag{11.B.1}$$

†Computed by Alvin Ng, EECS Dept., University of Illinois, Chicago, 1994.

where

m = order of the model
a_i = ith element of the AR parameter vector (alpha)

The true AR parameters as have been used (unknown to the neural network) to generate $x(n)$ are:

$$a_1 = 1.15$$
$$a_2 = 0.17$$
$$a_3 = -0.34$$
$$a_4 = -0.01$$
$$a_5 = 0.01$$

As in Sec. 6.D, an estimate of $\hat{a}_i$ of a_i is sought to minimize an MSE (mean square error) term given by

$$\text{MSE} = \hat{E}\left[e^2(n)\right] = \frac{1}{N}\sum_{i=1}^{N} e^2(i) \tag{11.B.2}$$

which is the sample variance (over N sampling points) of the error $e_{(n)}$ defined as

$$e(n) = x(n) - \hat{x}(n) \tag{11.B.3}$$

where

$$\hat{x}(n) = \sum_{i=1}^{m} \hat{a}_i x(n-i) \tag{11.B.4}$$

$\hat{a}(i)$ being the estimated (identified) AR parameters as sought by the neural network, exactly (so far) as was the deterministic case of Sec. 6.D. We note that Eq. (11.B.4) can also be written in vector form as:

$$\begin{aligned} \hat{x}(n) &= \hat{a}^T x(n)\,; \quad \hat{a} = [\hat{a}_1 \cdots \hat{a}_m]^T \\ \hat{x}_n &= [x(n-1) \cdots x(n-m)]^T \end{aligned} \tag{11.B.5}$$

T denoting transposition.

The procedure of stochastic training is described as the following: We define

$$E(n) = \gamma e^2(n) \tag{11.B.6}$$

where $\gamma = 0.5$, and $e(n)$ is the error energy. We subsequently update the weight vector (parameter vector estimate) $\hat{a}(n)$ of Eq. (11.B.5) by an uploadable $\Delta\tilde{a}$ given by

$$\Delta\hat{a}(n) = \rho T \cdot \tanh(r) \tag{11.B.7}$$

where

$$r = \text{random number from uniform distribution}$$

$$T = \text{temperature}$$

$$\rho = \text{learning rate coefficient} = 0.001$$

For this purpose we use a Cauchy procedure for simulated annealing. Since the Cauchy procedure may yield very large $\Delta\hat{a}$ which may cause the network to get stuck, $\Delta\hat{a}$ is modified as

$$\Delta\hat{a}_{\text{modified}} = -M + \frac{2m}{1 + \exp\left(-\dfrac{\Delta\hat{a}}{M}\right)} \tag{11.B.8}$$

where M is the hard limited value in which $-M \leq \Delta\hat{a}_{\text{modified}} \leq M$. Now, recalculate e, $e(n)$ with the new weight (parameter estimate) vector using Eqs. (11.B.3) to (11.B.6). If the error is reduced, the parameter estimate has been improved and accepts the new weight. If not, find the probability $P(\Delta e)$ of accepting this new weight from Cauchy distribution and also selected a random number r from a uniform distribution and compare this number with $P(\Delta e)$ which is defined as:

$$p(\Delta e) = \frac{T}{T^2 + \Delta^2 e} \tag{11.B.9}$$

where T is an equivalent (hypothetical) temperature value. If $P(\Delta e)$ is less than r, the network still accepts this worsening performance. Otherwise, restore the old weight (parameter estimate). Perform this process for each weight element. The temperature t should be decreased gradually to ensure convergence according to a temperature reduction algorithm:

$$T = \frac{T_o}{\log(1 + k)} \tag{11.B.10}$$

where $T_o = 200°$.

This weight updating is continued until the mean square error (MSE) is small enough, say MSE < 0.1. Then the network should stop.

The flow diagram of stochastic training is shown in Fig. 11.B.1.

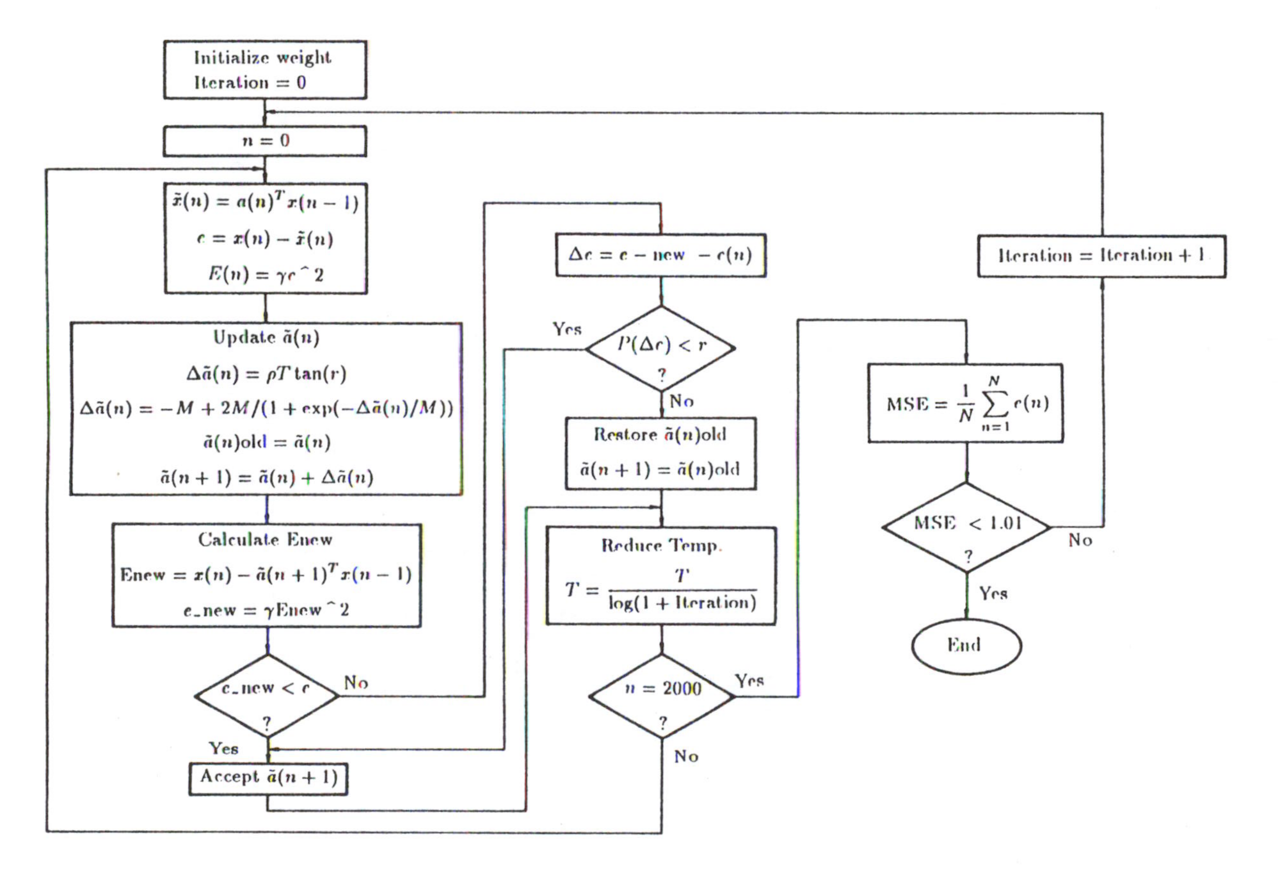

Fig. 11.B.1. Flow diagram of stochastic training.

11.B.2. *Program printout* (written in MATLAB® — see also Sec. 6.D)

```
   MATLAB FILE


   APPROACH:    STOCHASTIC TRAINING.


w1 = rand(5,1)/5;
delw = 0;
f1 = fopen('BP2.error','w');
fprintf(f1,'This is the training parameter of each iteration using Stochastic Traininga\r
fprintf(f1,'------------------------------------------------------------------------\r

gamma = 0.5;
rho = 0.001;
BOLTZ_CONST = 8.617/100;
TEMP = 300;      % INITIAL TEMPERATURE
M = 0.0006;        % HARD LIMIT ON THE AMPLITUDE OF delw

n = 5; ITERATION=2000;
Xpad = zeros(ITERATION+n,1);
Xpad(n+1:n+ITERATION) = X;
Xest = zeros(ITERATION+n,1);
error = zeros(size(X));
MSE = zeros(2000,1);
count = 1;
T = TEMP;
        for loop=1:40000
            for i=n+1:n+ITERATION
                xt = Xpad(i-n:i-1) ;
                dt = Xpad(i);
                E = dt - w1'*xt;
             error(i-n) = E;
                e = gamma*(E^2);
                    w1old = w1;
                % CAUCHY DISTRIBUTION OF WEIGHT CHANGE
                    delw = rho*T*tan(rand(5,1)*180/pi);
                % MODIFIED ON DELTA OMAGE delw
                    delw = -M + 2*M./(1+exp(-delw/M));
                    w1 = w1 + delw;
                    Enew = dt - w1'*xt;
                    enew = gamma*(Enew^2);
                    if enew < e
                       % DO NOTHING, KEEP delta_w
                    else
                       delE = enew - e;
                    %  if boltzmann(delE,BOLTZ_CONST,T) > rand(1)
                       if cauchy(delE,T) > rand(1)
                          % DO NOTHING, ACCEPT THIS WORSENING OF PERFORMANCE
                       else
                          % RESTORE OLD WEIGHT
                          w1 = w1old;
                       end
                    end
                % TEMPERATURE REDUCTION
                T = TEMP/log10(1+loop);
 if round(i/100)*100 == i
 %  disp([e  fliplr(w1')]);
 end
            end
 if  round(loop/10)*10 == loop
      MSE(loop) = (error'*error)/ITERATION;
```

```
     fprintf(f1,'\n%4d\t%7.4f\t%7.4f\t%7.4f\t%7.4f\t%7.4f',[loop  fliplr(w1')]);
     fprintf(f1,'\n  \t      MEANSQUARE ERROR = %6.4f\n\n',MSE(loop));
 end
        end

fclose(f1);
```

11.B.3. *Estimated parameter set at each iteration (using stochastic training)*

```
------------------------------------------------------------------------

  10     0.7769  0.2582  0.0991 -0.1246 -0.0452
               MEANSQUARE ERROR = 1.2805

  20     0.9435  0.1915  0.0005 -0.1217 -0.0303
               MEANSQUARE ERROR = 1.0878

  30     1.0092  0.1842 -0.0735 -0.1435  0.0141
               MEANSQUARE ERROR = 1.0704

  40 -   1.0778  0.1761 -0.1427 -0.1796  0.0557
               MEANSQUARE ERROR = 1.0488

  50     1.0555  0.1559 -0.1420 -0.1342  0.0406
               MEANSQUARE ERROR = 1.0482

  60     1.1062  0.1371 -0.2177 -0.0669  0.0380
               MEANSQUARE ERROR = 1.0309

  70     1.1142  0.0982 -0.2049 -0.0733  0.0226
               MEANSQUARE ERROR = 1.0443

  80     1.0693  0.1918 -0.2115 -0.0955  0.0194
               MEANSQUARE ERROR = 1.0350

  90     1.0444  0.2316 -0.1684 -0.1929  0.0748
               MEANSQUARE ERROR = 1.0478

 100     1.0707  0.2436 -0.1868 -0.1585 -0.0101
               MEANSQUARE ERROR = 1.0578

 110     1.0327  0.2228 -0.1414 -0.1915  0.0620
               MEANSQUARE ERROR = 1.0687

 120     0.9815  0.2631 -0.1140 -0.1457 -0.0076
               MEANSQUARE ERROR = 1.0609

 130     0.9532  0.3169 -0.1408 -0.2056  0.0534
               MEANSQUARE ERROR = 1.0629

 140     0.9414  0.3904 -0.2188 -0.1987  0.0953
               MEANSQUARE ERROR = 1.0749

 150     0.9473  0.4501 -0.2605 -0.2085  0.0540
               MEANSQUARE ERROR = 1.0739
```

```
160       0.9440  0.4304 -0.3314 -0.1286  0.0674
                MEANSQUARE ERROR = 1.0696

170       0.9992  0.4072 -0.3287 -0.0955 -0.0094
                MEANSQUARE ERROR = 1.0562

180       0.9527  0.4030 -0.3165 -0.0485 -0.0233
                MEANSQUARE ERROR = 1.0637

190       1.0022  0.3541 -0.2902 -0.0899  0.0154
                MEANSQUARE ERROR = 1.0496

200       1.0391  0.3314 -0.3456 -0.0738  0.0173
                MEANSQUARE ERROR = 1.0439

210       1.0158  0.3563 -0.2567 -0.1330  0.0024
                MEANSQUARE ERROR = 1.0412

220       1.0215  0.3571 -0.2440 -0.1683  0.0411
                MEANSQUARE ERROR = 1.0723

230       0.9804  0.3410 -0.2095 -0.1265  0.0126
                MEANSQUARE ERROR = 1.0526

240       1.0158  0.2938 -0.2264 -0.1370  0.0465
                MEANSQUARE ERROR = 1.0550

250       0.9579  0.2619 -0.1867 -0.0241 -0.0235
                MEANSQUARE ERROR = 1.0796

260       0.9859  0.2968 -0.2073 -0.0218 -0.0710
                MEANSQUARE ERROR = 1.0576

270       0.9902  0.2467 -0.1936 -0.0358 -0.0375
                MEANSQUARE ERROR = 1.0499

280       1.0531  0.2162 -0.2085 -0.0487 -0.0420
                MEANSQUARE ERROR = 1.0355

290       1.0592  0.1586 -0.1366 -0.0459 -0.0364
                MEANSQUARE ERROR = 1.0540

300       0.9678  0.1901 -0.0800 -0.0669 -0.0412
                MEANSQUARE ERROR = 1.0757

310       1.0208  0.1913 -0.1121 -0.0686 -0.0484
                MEANSQUARE ERROR = 1.0508

320       1.1222  0.1262 -0.1008 -0.1095 -0.0545
                MEANSQUARE ERROR = 1.0523
```

```
1810     1.0699  0.3708 -0.4338 -0.0244  0.0053
                MEANSQUARE ERROR = 1.0498

1820     1.1676  0.2528 -0.4968  0.0365  0.0249
                MEANSQUARE ERROR = 1.0458

1830     1.1143  0.2332 -0.3752  0.0477 -0.0215
                MEANSQUARE ERROR = 1.0336

1840     1.1391  0.1976 -0.4625  0.0816  0.0328
                MEANSQUARE ERROR = 1.0522

1850     1.0724  0.2908 -0.4057 -0.0012  0.0255
                MEANSQUARE ERROR = 1.0376

1860     1.0807  0.2777 -0.3864  0.0646 -0.0581
                MEANSQUARE ERROR = 1.0484

1870     1.0253  0.2623 -0.3482  0.0490 -0.0292
                MEANSQUARE ERROR = 1.0564

1880     1.0206  0.3282 -0.3571 -0.0071  0.0084
                MEANSQUARE ERROR = 1.0472

1890     1.0105  0.3136 -0.3425 -0.0476  0.0395
                MEANSQUARE ERROR = 1.0438

1900     1.0307  0.3859 -0.4360  0.0032 -0.0140
                MEANSQUARE ERROR = 1.0498

1910     1.0271  0.3962 -0.4764  0.0444 -0.0090
                MEANSQUARE ERROR = 1.0551

1920     1.0156  0.4603 -0.4436  0.0538 -0.1175
                MEANSQUARE ERROR = 1.1041

1930     0.9702  0.4214 -0.4306  0.0619 -0.0482
                MEANSQUARE ERROR = 1.0726

1940     0.9359  0.3811 -0.3517  0.0656 -0.0476
                MEANSQUARE ERROR = 1.0999

1950     1.0105  0.3883 -0.3834  0.1021 -0.1136
                MEANSQUARE ERROR = 1.0827

1960     1.0255  0.3171 -0.3448  0.0288 -0.0547
                MEANSQUARE ERROR = 1.0434

1970     1.0705  0.2556 -0.4278  0.1039 -0.0429
                MEANSQUARE ERROR = 1.0395
```

Observe that the convergence of the stochastic algorithm of the present case study is considerably slower than the convergence of its deterministic parallel of Sec. 4.A. This can be expected due to the randomness of the present search relative to the systematic nature of the deterministic algorithm. The main benefit of a stochastic algorithm is in its avoidance of local minima which does not occur in the present problem. In the present case we often seem to get very close to good estimates, only to be thrown off (by the randomness of the search) a bit later on. The next case study (Sec. 11.B below) will show some situations where a deterministic network is stuck at a local minimum and where, in certain situations, the stochastic network overcomes that difficulty.

Chapter 12

Recurrent (Time Cycling) Back Propagation Networks

12.1. Recurrent/Discrete Time Networks

A recurrent structure can be introduced into back propagation neural networks by feeding back the network's output to the input after an epoch of learning has been completed. This recurrent feature is in discrete steps (cycles) of weight computation. It was first proposed by Rumelhart *et al.* (1986) and subsequently by Pineda (1988), Hecht–Nielson (1990) and by Hertz *et al.* (1991). This arrangement allows the employment of back propagation with a small number of hidden layers (and hence of weights) in a manner that effectively is equivalent to using m-times that many layers if m cycles of recurrent computation are employed [cf. Fausett, 1993].

A recurrent (time cycling) back propagation network is described in Fig. 12.1. The delay elements (D in Fig. 12.1) in the feedback loops separate between the time-steps (epochs, which usually correspond to single iterations). At end of

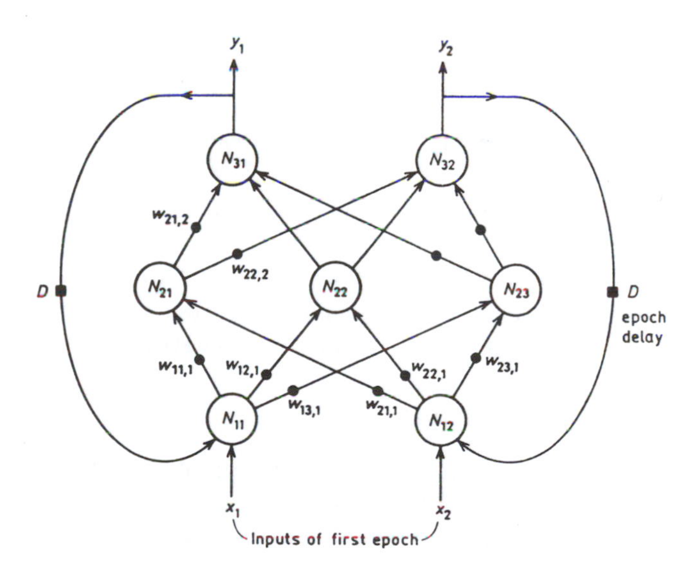

Fig. 12.1. A recurrent neural network structure.

the first epoch the outputs are fed back to the input. Alternatively, one may feed back the output-errors alone at the end of each epoch, to serve as inputs for the next epoch.

The network of Fig. 12.1 receives inputs x_1 and x_2 at various time steps of one complete sequence (set) that consititutes the first epoch (cycle). The weights are calculated as in conventional back-propagation networks and totalled over all time steps of an epoch with no actual adjustment of weights until the end of that epoch. At each time step the outputs y_1 and y_2 are fed back to be employed as the inputs for the next time step. At the end of one complete scan of all inputs, a next epoch is started with a new complete scan of the same inputs and time steps as in the previous epoch. When the number of inputs differs from the number of outputs, then the structure of Fig. 12.2 may be employed.

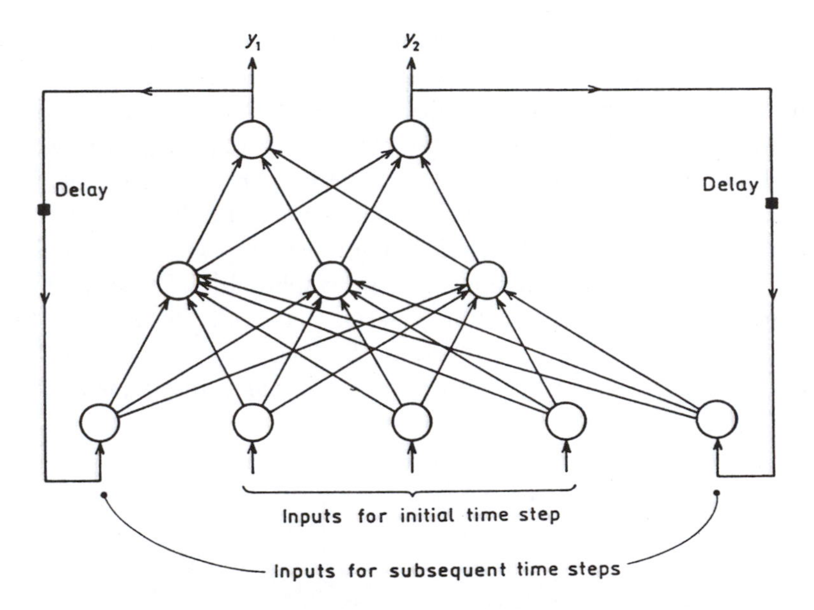

Fig. 12.2. A recurrent neural network structure (3 inputs/2 output).

Both structures in Figs. 12.1 and 12.2 are equivalent to a structure where the basic networks (except for the feedback from one time step to the other) is repeated m-times, to account for the time steps in the recurrent structure. See Fig. 12.3.

12.2. Fully Recurrent Networks

Fully recurrent networks are similar to the networks of Sec. 12.1 except that each layer feeds back to each preceding layer, as in Fig. 12.4 (rather than feeding back from the output of a n-layer network to the input of the network, as in Sec. 12.1). Now the output at each epoch becomes an input to a recurrent neuron at the next epoch.

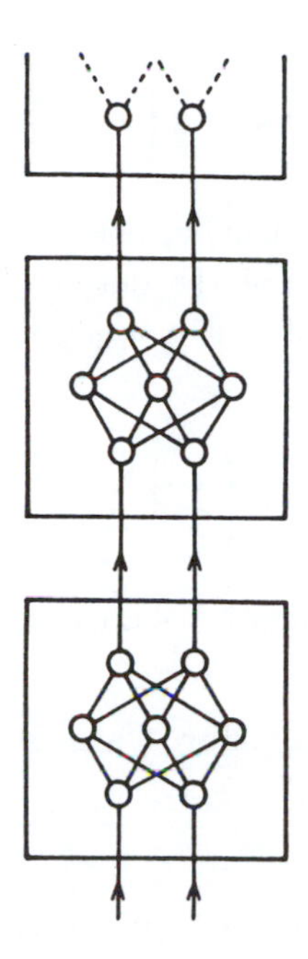

Fig. 12.3. A non-recurrent equivalent of the recurrent structures of Figs. 12.1, 12.2.

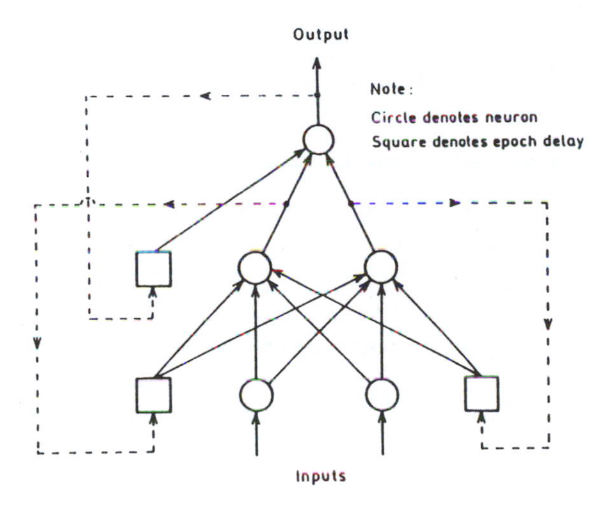

Fig. 12.4. A Fully Recurrent Back-Propagation Network.

12.3. Continuously Recurrent Back Propagation Networks

A continuously recurrent back propagation based neural network employs the same structure as in Figs. 12.1 and 12.2 but recurrency is repeated over infinitesimally small time interval. Hence, recurrency obeys a differential equation progression as in continuous Hopfield networks, namely

$$\tau \frac{dy_i}{dt} = -y_i + g\left(x_i + \sum_j w_{ij} v_j\right) \quad (12.1)$$

where τ is a time constant coefficient, x_i being the external input, $g(\cdots)$ denoting an activation function, y_i denoting the output and v_j being the outputs of the hidden layers neurons. For stability it is required that at least one stable solution of Eq. (12.1) exists, namely that

$$y_i = g\left(x_i + \sum_j w_{ij} v_j\right) \quad (12.2)$$

The Case Study of Sec. 6D illustrates a recurrent back-propagation algorithm.

12.A. Recurrent Back Propagation Case Study: Character Recognition*

12.A.1. *Introduction*

The present case study s concerned with solving a simple character recognition problem using a recurrent back propagation neural network. The task is to teach the neural network to recognize 3 characters, that is, to map them to respective pairs {0,1}, {1,0} and {1,1}. The network should also produce a special error signal 0,0 in response to any other character.

12.A.2. *Design of neural network*

Structure: The neural network consists of three layers with 2 neurons each, one output layer and two hidden layers. There are 36 regular inputs to the network and 2 inputs that are connected to the 2 output errors. Thus, in total there are 38 inputs to the neural network. The neural network is as in Section 6.A, except that it is a recurrent network, such that its outputs y1 and y2 are fed back as additional inputs at the end of each iteration. Bias terms (equal to 1) with trainable weights are also included in the network structure. The structural diagram of our neural network is given in Fig. 12.A.1.

(a) **Dataset Design:** The neural network is designed to recognize characters 'A', 'B' and 'C'. To train the network to produce error signal we will use another 6 characters: 'D', 'E', 'F', 'G', 'H' and 'I'. To check whether the network has learned to recognize errors we will use characters 'X', 'Y' and 'Z'. Note that we are interested in checking the response of the network to errors on the characters which were not involved in the training procedure. The characters to be recognized are given on a 6×6 grid. Each of the 36 pixels is set to either 0 or 1. Corresponding 6×6 matrices are as follows:

---

*Computed by Maxim Kolesnikov, ECE Dept., University of Illinois, Chicago, 2006.

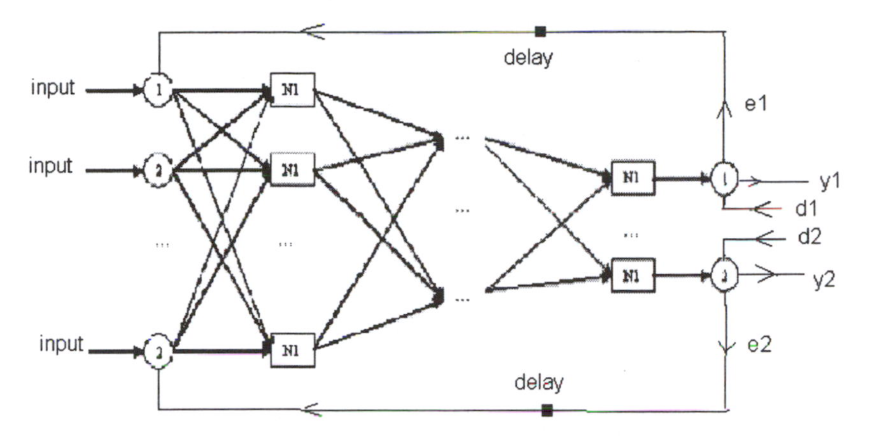

Fig. 12.A.1: Recurrent back-propagation neural network.

A: 001100	B: 111110	C: 011111
010010	100001	100000
100001	111110	100000
111111	100001	100000
100001	100001	100000
100001	111110	011111
D: 111110	E: 111111	F: 111111
100001	100000	100000
100001	111111	111111
100001	100000	100000
100001	100000	100000
111110	111111	100000
G: 011111	H: 100001	I: 001110
100000	100001	000100
100000	111111	000100
101111	100001	000100
100001	100001	000100
011111	100001	001110
X: 100001	Y: 010001	Z: 111111
010010	001010	000010
001100	000100	000100
001100	000100	001000
010010	000100	010000
100001	000100	111111

(b) **Setting of Weights:** Back propagation learning was used to solve the problem. The goal of this algorithm is to minimize the error-energy at the output layer. Weight setting is as in regular Back-Propagation, Section 6.2 of Chapter 6 above.

A source code for this case study (written in C++) is given in Sect. 12.A.5.

12.A.3. *Results*

(a) Training Mode

To train the network to recognize the above characters we applied corresponding 6×6 grids in the form of 1×36 vectors to the input of the network. Additional two inputs were initially set equal to zero and in the course of the training procedure were set equal to the current output error. A character was considered recognized if both outputs of the network were no more than 0.1 off their respective desired values. The initial learning rate η was experimentally set at 1.5 and was decreased by a factor of 2 after each 100th iteration. Just as in regular back propagation (Sect. 6.A), after each 400th iteration we reset the learning rate to its initial value, in order to prevent the learning process from getting stuck at a local minimum. Then after about 3000 iterations we were able to correctly recognize all datasets. We, however, continued until 5000 iterations were completed to make sure that the energy-error value cannot be lowered even further. At this point we obtained:

TRAINING VECTOR 0: [0.0296153 0.95788] — RECOGNIZED —
TRAINING VECTOR 1: [0.963354 2.83491e-06] — RECOGNIZED —
TRAINING VECTOR 2: [0.962479 0.998554] — RECOGNIZED —
TRAINING VECTOR 3: [0.0162449 0.0149129] — RECOGNIZED —
TRAINING VECTOR 4: [0.0162506 0.0149274] — RECOGNIZED —
TRAINING VECTOR 5: [0.0161561 0.014852] — RECOGNIZED —
TRAINING VECTOR 6: [0.0168284 0.0153119] — RECOGNIZED —
TRAINING VECTOR 7: [0.016117 0.0148073] — RECOGNIZED —
TRAINING VECTOR 8: [0.016294 0.0149248] — RECOGNIZED —

Training vectors $0, 1, \ldots, 8$ in these log entries correspond to the characters 'A', 'B', ..., 'I'.

3(b) Recognition Results (test runs)

Error Detection: To check error detection performance, we saved the obtained weights into a data file, modified the datasets in the program replacing the characters 'G', 'H' and 'I' (training vectors 6, 7 and 8) by the characters 'X', 'Y' and 'Z'. Then we ran the program, loaded the previously saved weights from the data file and applied the input to the network. Note that we performed no further training. We got the following results:

TRAINING VECTOR 6: [0.00599388 0.00745234] — RECOGNIZED —
TRAINING VECTOR 7: [0.0123415 0.00887678] — RECOGNIZED —
TRAINING VECTOR 8: [0.0433571 0.00461456] — RECOGNIZED —

All three characters were successfully mapped to error signal $\{0, 0\}$.

Robustness: To investigate how robust our neural network was, we added some noise to the input and got the following results. In the case of 1-bit distortion (out of 36 bits) the recognition rates were:

TRAINING SET 0: 18/38 recognitions (47.3684%)
TRAINING SET 1: 37/38 recognitions (97.3684%)
TRAINING SET 2: 37/38 recognitions (97.3684%)
TRAINING SET 3: 5/38 recognitions (13.1579%)
TRAINING SET 4: 5/38 recognitions (13.1579%)
TRAINING SET 5: 5/38 recognitions (13.1579%)
TRAINING SET 6: 6/38 recognitions (15.7895%)
TRAINING SET 7: 5/38 recognitions (13.1579%)
TRAINING SET 8: 6/38 recognitions (15.7895%)

With 2 error bits per character, performance was even worse.

12.A.4. *Discussion and conclusions*

We were able to train our neural network so that it successfully recognizes the three given characters and at the same time is able to classify other characters as errors. However, the results are not spectacular for the distorted input datasets. Characters 'A', 'B' and 'C', that our network was trained on, were successfully recognized with 1 and 2 bit distortions (with the possible exception of character 'A' but it could be improved by increasing the number of iterations). But recognition of the 'rest of the world' characters was not great.

Comparing this result with the result achieved using pure back propagation, we can see that for this particular problem, if noise bits were added to the data, recurrency worsened the recognition performance results as compared with regular (non-recurrent) Back-Propagation. Also, due to the introduction of recurrent inputs we had to increase the total number of inputs by two. This resulted in the increased number of weights in the network and, therefore, in somewhat slower learning.

12.A.5. *Source code (C++)*

```
/*

*/
#include<cmath>
#include<iostream>
#include<fstream>
using namespace std;
#define N_DATASETS 9
#define N_INPUTS 38
#define N_OUTPUTS 2
#define N_LAYERS 3
// {# inputs, # of neurons in L1, # of neurons in L2, # of neurons in
// L3}
short conf[4] = {N_INPUTS, 2, 2, N_OUTPUTS};
// According to the number of layers double **w[3], *z[3], *y[3], *Fi[3], eta; ofstream
ErrorFile("error.txt", ios::out);
// 3 training sets; inputs 36 and 37 (starting from 0) will be used
// for feeding back the output error bool dataset[N_DATASETS][N_INPUTS] = {

{ 0, 0, 1, 1, 0, 0,       // 'A'
  0, 1, 0, 0, 1, 0,
  1, 0, 0, 0, 0, 1,
  1, 1, 1, 1, 1, 1,
  1, 0, 0, 0, 0, 1,
  1, 0, 0, 0, 0, 1, 0, 0},
{ 1, 1, 1, 1, 1, 0,       // 'B'
  9

  1, 0, 0, 0, 0, 1,
  1, 1, 1, 1, 1, 0,
  1, 0, 0, 0, 0, 1,
  1, 0, 0, 0, 0, 1,
  1, 1, 1, 1, 1, 0, 0, 0},
{ 0, 1, 1, 1, 1, 1,        // 'C'
  1, 0, 0, 0, 0, 0,
  1, 0, 0, 0, 0, 0,
  1, 0, 0, 0, 0, 0,
  1, 0, 0, 0, 0, 0,
  0, 1, 1, 1, 1, 1, 0, 0},
{ 1, 1, 1, 1, 1, 0,        // 'D'
  1, 0, 0, 0, 0, 1,
  1, 0, 0, 0, 0, 1,
  1, 0, 0, 0, 0, 1,
  1, 0, 0, 0, 0, 1,
  1, 1, 1, 1, 1, 0, 0, 0},
{ 1, 1, 1, 1, 1, 1,        // 'E'
  1, 0, 0, 0, 0, 0,
  1, 1, 1, 1, 1, 1,
  1, 0, 0, 0, 0, 0,
  1, 0, 0, 0, 0, 0,
  1, 1, 1, 1, 1, 1, 0, 0},
{ 1, 1, 1, 1, 1, 1,        // 'F'
  1, 0, 0, 0, 0, 0,
```

```
   1,  1,  1,  1,  1,  1,
   1,  0,  0,  0,  0,  0,
   1,  0,  0,  0,  0,  0,
   1,  0,  0,  0,  0,  0,  0,  0},
{  0,  1,  1,  1,  1,  1,         //  'G'
   1,  0,  0,  0,  0,  0,
   1,  0,  0,  0,  0,  0,
   1,  0,  1,  1,  1,  1,
   1,  0,  0,  0,  0,  1,
   0,  1,  1,  1,  1,  1,  0,  0},
{  1,  0,  0,  0,  0,  1,         //  'H'
   1,  0,  0,  0,  0,  1,
   1,  1,  1,  1,  1,  1,
  10

   1,  0,  0,  0,  0,  1,
   1,  0,  0,  0,  0,  1,
   1,  0,  0,  0,  0,  1,  0,  0},
{  0,  0,  1,  1,  1,  0,         //  'I'
   0,  0,  0,  1,  0,  0,
   0,  0,  0,  1,  0,  0,
   0,  0,  0,  1,  0,  0,
   0,  0,  0,  1,  0,  0,
   0,  0,  1,  1,  1,  0,  0,  0}
// Below are the datasets for checking "the rest of the world". They
// are not the  ones the NN was trained on.
/*
{  1,  0,  0,  0,  0,  1,         //  'X'
   0,  1,  0,  0,  1,  0,
   0,  0,  1,  1,  0,  0,
   0,  0,  1,  1,  0,  0,
   0,  1,  0,  0,  1,  0,
   1,  0,  0,  0,  0,  1,  0,  0},
{  0,  1,  0,  0,  0,  1,         //  'Y'
   0,  0,  1,  0,  1,  0,
   0,  0,  0,  1,  0,  0,
   0,  0,  0,  1,  0,  0,
   0,  0,  0,  1,  0,  0,
   0,  0,  0,  1,  0,  0,  0,  0},
{  1,  1,  1,  1,  1,  1,         //  'Z'
   0,  0,  0,  0,  1,  0,
   0,  0,  0,  1,  0,  0,
   0,  0,  1,  0,  0,  0,
   0,  1,  0,  0,  0,  0,
   1,  1,  1,  1,  1,  1,  0,  0}*/
},
datatrue[N_DATASETS][N_OUTPUTS] = {{0,1}, {1,0}, {1,1},
{0,0}, {0,0}, {0,0}, {0,0}, {0,0}, {0,0}};
// Memory allocation and initialization function void MemAllocAndInit(char S)
{
if(S == 'A')
for(int i = 0; i < N_LAYERS; i++)
11

{
```

```
w[i] = new double*[conf[i + 1]]; z[i] = new double[conf[i + 1]];
y[i] = new double[conf[i + 1]]; Fi[i] = new double[conf[i + 1]];
for(int j = 0; j < conf[i + 1]; j++)
{
}
}
w[i][j] = new double[conf[i] + 1];
// Initializing in the range (-0.5;0.5) (including bias
// weight)
for(int k = 0; k <= conf[i]; k++)
w[i][j][k] = rand()/(double)RAND_MAX - 0.5;
if(S == 'D')
{
for(int i = 0; i < N_LAYERS; i++)
{
}
for(int j = 0; j < conf[i + 1]; j++)
delete[] w[i][j];
delete[] w[i], z[i], y[i], Fi[i];
}
}
ErrorFile.close();
// Activation function double FNL(double z)
{
}
double y;
y = 1. / (1. + exp(-z));
return y;
// Applying input
void ApplyInput(short sn)
{
double input;
12

// Counting layers
for(short i = 0; i < N_LAYERS; i++)
// Counting neurons in each layer for(short j = 0; j < conf[i + 1]; j++)
{
z[i][j] = 0.;
// Counting input to each layer (= # of neurons in the previous
// layer)
for(short k = 0; k < conf[i]; k++)
{
// If the layer is not the first one if(i)
input = y[i - 1][k];
else
input = dataset[sn][k];
z[i][j] += w[i][j][k] * input;
}
}
}
z[i][j] += w[i][j][conf[i]];          // Bias term y[i][j] = FNL(z[i][j]);
// Training function, tr - # of runs void Train(int tr)
{
short i, j, k, m, sn;
```

```
double eta, prev_output, multiple3, SqErr, eta0;
// Starting learning rate eta0 = 1.5;
eta = eta0;
// Going through all tr training runs for(m = 0; m < tr; m++)
{
SqErr = 0.;
// Each training run consists of runs through each training set for(sn = 0;
sn < N_DATASETS; sn++)
{
13

ApplyInput(sn);
// Counting the layers down
for(i = N_LAYERS - 1; i >= 0; i--)
// Counting neurons in the layer for(j = 0; j < conf[i + 1]; j++)
{
if(i == 2)     // If it is the output layer multiple3 = datatrue[sn][j] - y[i][j];
else
{
}
multiple3 = 0.;
// Counting neurons in the following layer for(k = 0; k < conf[i + 2]; k++)
multiple3 += Fi[i + 1][k] * w[i + 1][k][j];
Fi[i][j] = y[i][j] * (1 - y[i][j]) * multiple3;
// Counting weights in the neuron
// (neurons in the previous layer)
for(k = 0; k < conf[i]; k++)
{
{
switch(k)
{
case 36:
if(i) // If it is not a first layer prev_output = y[i - 1][k];
else
prev_output = y[N_LAYERS - 1][0] - datatrue[sn][0];
break;
case 37:
prev_output = y[N_LAYERS - 1][1] - datatrue[sn][1];
break;
default:
prev_output = dataset[sn][k];
}
}
}
w[i][j][k] += eta * Fi[i][j] * prev_output;
14

}
// Bias weight correction w[i][j][conf[i]] += eta * Fi[i][j];
}
SqErr += pow((y[N_LAYERS - 1][0] - datatrue[sn][0]), 2) +
pow((y[N_LAYERS - 1][1] - datatrue[sn][1]), 2);
}
}
ErrorFile << 0.5 * SqErr << endl;
```

```
// Decrease learning rate every 100th iteration if(!(m % 100))
eta /= 2.;
// Go back to original learning rate every 400th iteration if(!(m % 400))
eta = eta0;
// Prints complete information about the network void PrintInfo(void)
{
// Counting layers
for(short i = 0; i < N_LAYERS; i++)
{
cout << "LAYER " << i << endl;
// Counting neurons in each layer for(short j = 0; j < conf[i + 1]; j++)
{
cout << "NEURON " << j << endl;
// Counting input to each layer (= # of neurons in the previous
// layer)
for(short k = 0; k < conf[i]; k++)
cout << "w[" << i << "][" << j << "][" << k << "]="
<< w[i][j][k] << ' ';
cout << "w[" << i << "][" << j << "][BIAS]="
<< w[i][j][conf[i]] << ' ' << endl;
cout << "z[" << i << "][" << j << "]=" << z[i][j] << endl;
cout << "y[" << i << "][" << j << "]=" << y[i][j] << endl;
}
}
15

}
// Prints the output of the network void PrintOutput(void)
{
// Counting number of datasets
for(short sn = 0; sn < N_DATASETS; sn++)
{
}
}
ApplyInput(sn);
cout << "TRAINING SET " << sn << ": [ ";
// Counting neurons in the output layer for(short j = 0; j < conf[3]; j++)
cout << y[N_LAYERS - 1][j] << ' ';
cout << "] ";
if(y[N_LAYERS - 1][0] > (datatrue[sn][0] - 0.1)
&& y[N_LAYERS - 1][0] < (datatrue[sn][0] + 0.1)
&& y[N_LAYERS - 1][1] > (datatrue[sn][1] - 0.1)
&& y[N_LAYERS - 1][1] < (datatrue[sn][1] + 0.1))
cout << "--- RECOGNIZED ---";
else
cout << "--- NOT RECOGNIZED ---";
cout << endl;
// Loads weihts from a file void LoadWeights(void)
{
double in;
ifstream file("weights.txt", ios::in);
// Counting layers
for(short i = 0; i < N_LAYERS; i++)
// Counting neurons in each layer for(short j = 0; j < conf[i + 1]; j++)
// Counting input to each layer (= # of neurons in the previous
```

```
// layer)
for(short k = 0; k <= conf[i]; k++)
{
16

}
file >> in;
w[i][j][k] = in;
}
file.close();
// Saves weiths to a file void SaveWeights(void)
{
}
ofstream file("weights.txt", ios::out);
// Counting layers
for(short i = 0; i < N_LAYERS; i++)
// Counting neurons in each layer for(short j = 0; j < conf[i + 1]; j++)
// Counting input to each layer (= # of neurons in the previous
// layer)
for(short k = 0; k <= conf[i]; k++)
file << w[i][j][k] << endl;
file.close();
// Gathers recognition statistics for 1 and 2 false bit cases void
GatherStatistics(void)
{
short sn, j, k, TotalCases;
int cou;
cout << "WITH 1 FALSE BIT PER CHARACTER:" << endl; TotalCases = conf[0];
// Looking at each dataset
for(sn = 0; sn < N_DATASETS; sn++)
{
cou = 0;
// Looking at each bit in a dataset for(j = 0; j < conf[0]; j++)
{
if(dataset[sn][j])
dataset[sn][j] = 0;
17

}
else
 dataset[sn][j] = 1; ApplyInput(sn);
if(y[N_LAYERS - 1][0] > (datatrue[sn][0] - 0.1)
&& y[N_LAYERS - 1][0] < (datatrue[sn][0] + 0.1)
&& y[N_LAYERS - 1][1] > (datatrue[sn][1] - 0.1)
&& y[N_LAYERS - 1][1] < (datatrue[sn][1] + 0.1))
cou++;
// Switching back if(dataset[sn][j])
dataset[sn][j] = 0;
else
dataset[sn][j] = 1;
}
cout << "TRAINING SET " << sn << ": " << cou << '/' << TotalCases
<< " recognitions (" << (double)cou / TotalCases * 100. << "%)" << endl;
cout << "WITH 2 FALSE BITS PER CHARACTER:" << endl;
TotalCases = conf[0] * (conf[0] - 1);
```

```
// Looking at each dataset
for(sn = 0; sn < N_DATASETS; sn++)
{
cou = 0;
// Looking at each bit in a dataset for(j = 0; j < conf[0]; j++)
for(k = 0; k < conf[0]; k++)
{
if(j == k)
continue;
if(dataset[sn][j])
dataset[sn][j] = 0;
else
dataset[sn][j] = 1;
if(dataset[sn][k])
dataset[sn][k] = 0;
else
dataset[sn][k] = 1;
18

}
ApplyInput(sn);
if(y[N_LAYERS - 1][0] > (datatrue[sn][0] - 0.1)
&& y[N_LAYERS - 1][0] < (datatrue[sn][0] + 0.1)
&& y[N_LAYERS - 1][1] > (datatrue[sn][1] - 0.1)
&& y[N_LAYERS - 1][1] < (datatrue[sn][1] + 0.1))
cou++;
if(dataset[sn][j]) // Switching back dataset[sn][j] = 0;
else
dataset[sn][j] = 1;
if(dataset[sn][k])
dataset[sn][k] = 0;
else
dataset[sn][k] = 1;
}
}
cout << "TRAINING SET " << sn << ": " << cou << '/' << TotalCases
<< " recognitions (" << (double)cou / TotalCases * 100. << "%)" << endl;
// Entry point: main menu int main(void)
{
short ch;
int x;
MemAllocAndInit('A');
do
{
cout << "MENU" << endl;
cout << "1. Apply input and print parameters" << endl;
cout << "2. Apply input (all training sets) and print output" << endl;
cout << "3. Train network" << endl; cout << "4. Load weights" << endl;
cout << "5. Save weights" << endl;
cout << "6. Gather recognition statistics" << endl;
cout << "0. Exit" << endl;
19

cout << "Your choice: ";
cin >> ch; cout << endl; switch(ch)
```

```
{
case 1: cout << "Enter set number: ";
cin >> x; ApplyInput(x); PrintInfo(); break;
case 2: PrintOutput();
break;
case 3: cout << "How many training runs?: ";
cin >> x; Train(x); break;
case 4: LoadWeights();
break;
case 5: SaveWeights();
break;
case 6: GatherStatistics();
break;
case 0: MemAllocAndInit('D');
return 0;
}
}
cout << endl;
cin.get();
cout << "Press ENTER to continue..." << endl;
cin.get();
}
while(ch);
20
```

Chapter 13

Large Scale Memory Storage and Retrieval (LAMSTAR) Network

13.0. Motivation

The neural network discussed in the present section is an artificial neural network for large scale memory storage and retrieval of information [Graupe and Kordylewski, 1996a,b]. This network attempts to imitate, in a gross manner, processes of the human central nervous system (CNS), concerning storage and retrieval of patterns, impressions and sensed observations, including processes of forgetting and of recall. It attempts to achieve this without contradicting findings from physiological and psychological observations, at least in an input/output manner. Furthermore, the LAMSTAR (LArge Memory STorage And Retrieval) model considered attempts to do so in a computationally efficient manner, using tools of neural networks from the previous sections, especially *SOM* (Self Organizing Map)-based network modules (similar to those of Sec. 8 above), combined with statistical decision tools. The LAMSTAR network is therefore *not a specific network but a system of networks* for storage, recognition, comparison and decision that in combination allow such storage and retrieval to be accomplished.

13.1. Basic Principles of the LAMSTAR Neural Network

The LAMSTAR neural network is specifically designed for application to retrieval, diagnosis, classification, prediction and decision problems which involve a very large number of categories. The resulting LAMSTAR (**LA**rge **M**emory **ST**orage **A**nd **R**etrieval) neural network [graupe, 1997, Graupe and Kordylewski, 1998] is designed to store and retrieve patterns in a computationally efficient manner, using tools of neural networks, especially Kohonen's SOM (Self Organizing Map)-based network modules [Kohonen, 1988], combined with statistical decision tools.

By its structure as described in Sec. 13.2, the LAMSTAR network is uniquely suited to deal with analytical and non-analytical problems where data are of many vastly different categories and where some categories may be missing, where data are both exact and fuzzy and where the vastness of data requires very fast algorithms

[Graupe, 1997, Graupe and Kordylewski, 1998]. These features are rare to find, especially when coming together, in other neural networks.

The LAMSTAR can be viewed as in intelligent expert system, where expert information is continuously being ranked for each case through learning and correlation. What is unique about the LAMSTAR network is its capability to deal with non-analytical data, which may be exact or fuzzy and where some categories may be missing. These characteristics are facilitated by the network's features of forgetting, interpolation and extrapolation. These allow the network to zoom out of stored information via forgetting and still being able to approximate forgotten information by extrapolation or interpolation. The LAMSTAR was specifically developed for application to problems involving very large memory that relates to many different categories (attributes), where some of the data is exact while other data are fuzzy and where (for a given problem) some data categories may occasionally be totally missing. Also, the LAMSTAR NN is insensitive to initialization and is doe not converge to local minima. Furthermore, in contrast to most Neural Networks (say, Back-Propagation as in Chapter 6), the LAMSTAR's unique weight structure makes it fully transparent, since its weights provide clear information on what is going on inside the network. Consequently, the network has been successfully applied to many decision, diagnosis and recognition problems in various fields.

The major principles of neural networks (NN's) are common to practically all NN approaches. Its elementary neural unit or cell (neuron) is the one employed in all NN's, as described in Chapters 2 and 4 of this text. Accordingly, if the p inputs into a given neuron (from other neurons or from sensors or transducers at the input to the whole or part of the whole network) at the j'th SOM layer are denoted as $x(ij); i = 1, 2, \ldots, p$, and if the (single) output of that neuron is denoted as y, then the neuron's output y satisfies;

$$y = f\left[\sum_{i=1}^{p} w_{ij} x_{ij}\right] \tag{13.1}$$

where $f[.]$ is a nonlinear function denoted as Activation Function, that can be considered as a (hard or soft) binary (or bipolar) switch, as in Chapter 4 above. The weights w_{ij} of Eq. (13.1) are the weights assigned to the neuron's inputs and whose setting is the learning action of the NN. Also, neural firing (producing of an output) is of all-or-nothing nature [McCulloch and Pitts, 1943]. For details of the setting of the storage weights (w_{ij}), see Secs. 13.2.2 and 13.2.6 below.

The WTA (Winner-Take-All) principle, as in Chapter 8, is employed [Kohonen, 1988], such that an output (firing) is produced ***only*** at the winning neuron, namely, at the output of the neuron whose storage weights w_{ij} are closest to vector $\underline{\mathbf{x}}(j)$ when a best-matching memory is sought at the j'th SOM module.

By using a link weights structure for its decision and browsing, the LAMSTAR network utilizes not just the stored memory values $w(ij)$ as in other neural networks, but also the interrelations these memories (Verbindungen, as Immanuel Kant called them [Ewing, 1938]) to the decision module and between the memories themselves. The LAMSTAR's understanding is thus based not just on memories (in terms of its storage weights) but also on relations between them, (in terms of link weights). These relations (link weights) are fundamental to its operation. By Hebb's Law [Hebb, 1949], interconnecting weights (link weights) adjust and serve to establish flow of neuronal signal traffic between groups of neurons, such that when a certain neuron fires very often in close time proximity (regarding a given situation/task), then the interconnecting link-weights (not the memory-storage weights) relating to that traffic, increase as compared to other interconnections [Graupe, 1997; Graupe and Lynn, 1970]. Indeed, link weights serve as Hebbian intersynaptic weights and adjust accordingly [Hebb, 1949]. These weights and their method of adjustment (according to flow of traffic in the interconnections) fit recent results from brain research [Levitan et al., 1997]. They are also responsible to the LAMSTAR's ability to interpolate/extrapolate and perform (with no re-programming or retraining) with incomplete dare sets.

13.2. Detailed Outline of the LAMSTAR Network

13.2.1. *Basic structural elements*

The basic storage modules of the LAMSTAR network are modified Kohonen SOM modules [Kohonen, 1988] of Chapter 8 that are Asociate-Memory-based WTA, in accordance to degree of proximity of **storage weights** in the BAM-sense to any **input subword** that is being considered per any given **input word** to the NN. In the LAMSTAR network the information is stored and processed via correlation links between individual neurons in separate SOM modules. Its ability to deal with a large number of categories is partly due to its use of simple calculation of **link weights** and by its use of **forgetting** features and features of recovery from forgetting. The link weights are the main engine of the network, connecting many layers of SOM modules such that the emphasis is on (co)relation of link weights between atoms of memory, not on the memory atoms (BAM weights of the SOM modules) themselves. In this manner, the design becomes closer to knowledge processing in the biological central nervous system than is the practice in most conventional artificial neural networks. The forgetting feature too, is a basic feature of biological networks whose efficiency depends on it, as is the ability to deal with incomplete data sets.

The **input word** is a coded real matrix $\boldsymbol{X}$ given by:

$$\underline{X} = \left[\underline{x}_1^T, \underline{x}_2^T, \ldots, \underline{x}_N^T\right]^T \tag{13.2}$$

where T denotes transposition., $\boldsymbol{x}_i^T$ being subvectors (subwords describing categories or attributes of the input word). Each subword $\underline{\boldsymbol{x}}_i$ is channeled to a corresponding i'th SOM module that stores data concerning the i'th category of the input word.

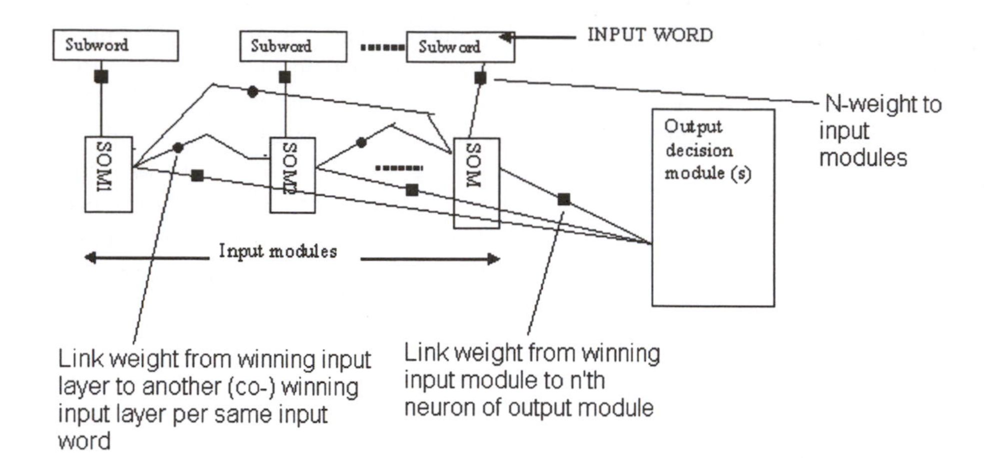

Fig. 13.1. A generalized LAMSTAR block-diagram.

Many input subwords (and similarly, many inputs to practically any other neural network approach) can be derived only after ***pre-processing***. This is the case in signal/image-processing problems, where only autoregressive or discrete spectral/wavelet parameters can serve as a subword rather than the signal itself.

Whereas in most SOM networks [Kohonen, 1988] all neurons of an SOM module are checked for proximity to a given input vector, in the LAMSTAR network only a finite group of p neurons may checked at a time due to the huge number of neurons involved (the large memory involved). The final set of p neurons is determined by link-weights (N_i) as shown in Fig. 13.1. However, if a given problem requires (by considerations of its quantization) only a small number of neurons in a given SOM storage module (namely, of possible states of an input subword), then all neurons in a given SOM module will be checked for possible storage and for subsequent selection of a winning neuron in that SOM module (layer) and N_i weights are *not* used. Consequently, if the number of quantization levels in an input subword is small, then the subword is channeled directly to all neurons in a predetermined SOM module (layer).

The main element of the LAMSTAR, which is its decision engine, is the array of link weights that interconnect neurons between input SOM layers and from all storage neurons of the inpout layers to the output (decision) layers. The inter-input-layer link weights are updated in accordance with traffic volume. The link weights to the output layers are updated by a reward/punishment process in accordance to success or failure of any decision, thus forming a learning process that is not limited to training data but continuous throughout running the LAMSTAR on a give problem. Weight-initialization is simple and unproblematic. Its feed-forward structure guarantees its stability, as is also discussed below. Details on the link weight adjustments and related topics are discussed in the sections below.

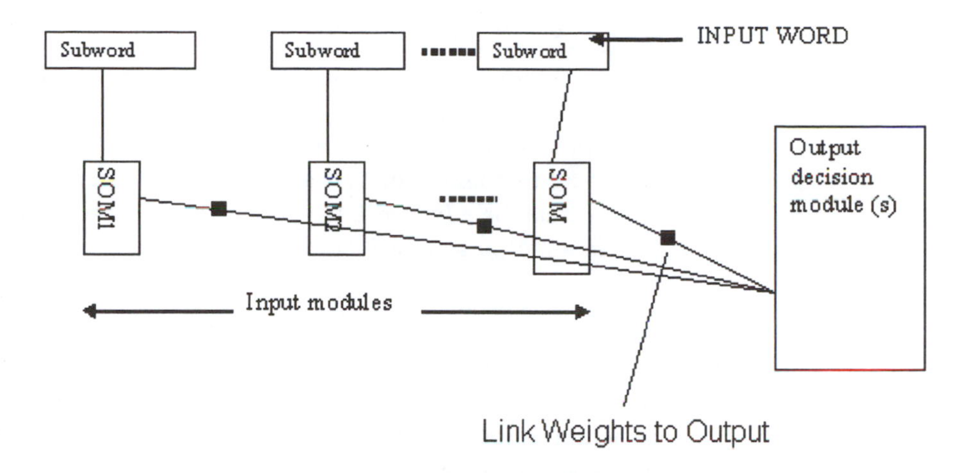

Fig. 13.2. The basic LAMSTAR architecture: simplified version for most applications.

Figure 13.1 gives a block-diagram of the complete and generalized of the LAMSTAR network. A more basic diagram, to be employed in most applications where the number of neurons per SOM layer is not huge, is given in Fig. 13.2. This design is a slight simplification of the generalized architecture. It is also employed in the case studies of Appendices 13.A and 13.B below. Only large browsing/retrieval cases should employ the complete design of Fig. 13.1. In the design of Fig. 13.2, the internal weights from one input layer to other input layers are omitted, as are the N_{ij} weights.

13.2.2. *Setting of storage weights and determination of winning neurons*

When a new input word is presented to the system during the training phase, the LAMSTAR network inspects all storage-weight vectors ($\boldsymbol{w}_i$) in SOM module i that corresponds to an input subword $\underline{\mathbf{x}}_i$ that is to be stored. If any stored pattern matches the input subword $\underline{\mathbf{x}}_i$ within a preset tolerance, it is declared as the **winning** neuron for that particularly observed input subword. A **winning** neuron is thus determined for each input based on the similarity between the input (vector $\boldsymbol{x}$ in Figs. 13.1 and 13.2) and a storage-weight vector $\underline{w}$ (stored information). For an input subword $\mathbf{x}_i$, the winning neuron is thus determined by minimizing a distance norm $\| * \|$, as follows:

$$d(j,j) = \|\underline{x}_j - \underline{w}_j\| \leq \|\underline{x}_j - \underline{w}_{k\neq j}\| \triangleq d(j,k) \qquad \forall\ k \tag{13.3}$$

13.2.3. *Adjustment of resolution in SOM modules*

Equation (13.3), which serves to determine the winning neuron, does not deal effectively with the resolution of close clusters/patterns. This may lead to degraded accuracy in the decision making process when decision depends on local and closely related patterns/clusters which lead to different diagnosis/decision. The local sensitivity of neuron in SOM modules can be adjusted by incorporating an adjustable maximal Hamming distance function $d_{\max}$ as in Eq. (13.4):

$$d_{\max} = \max[d(\underline{x}_i\, \underline{w}_i)]\,. \tag{13.4}$$

Consequently, if the number of subwords stored in a given neuron (of the appropriate module) exceeds a threshold value, then storage is divided into two adjacent storage neurons (i.e. a new-neighbor neuron is set) and dmax is reduced accordingly.

For fast adjustment of resolution, link weight to the output layer (as discussed in Sec. 13.2.3 below) can serve to adjust the resolution, such that storage in cells that yield a relatively high N_{ij} weights can be divided (say into 2 cells), while cells with low output link weights can be merged into the neighboring cells. This adjustment can be automatically or periodically changed when certain link weights increase or decrease relative to others over time (and considering the networks forgetting capability as in Sec. 13.3 below).

13.2.4. *Links between SOM modules and from SOM modules to output modules*

Information in the LAMSTAR system is encoded via correlation links $L_{i,j}$ (Figs. 13.1, 13.2) between individual neurons in different SOM modules. The LAMSTAR system does not create neurons for an entire input word. Instead, only selected subwords are stored in Associative-Memory-like manner in SOM modules (w weights), and correlations between subwords are stored in terms of creating/adjusting L-links ($L_{i,j}$ in Fig. 13.1) that connect neurons in different SOM modules. This allows the LAMSTAR network to be trained with partially incomplete data sets. The L-links are fundamental to allow interpolation and extrapolation of patterns (when a neuron in an SOM model does not correspond to an input subword but is highly linked to other modules serves as an interpolated estimate). We comment that the setting (updating) of Link Weights, as considered in this sub-section, applies to both link weights between **input-storage** (internal) SOM modules **AND also** link-weights from any storage SOM module and an **output module** (layer). **In many applications it is advisable to consider only links to ouput (decision) modules**.

Specifically, **link weight values** L are **set** (updated) such that for a given input word, after determining a **winning** k'th neuron in module i and a winning m'th neuron in module j, then the link weight $L_{i,j}^{k,m}$ is counted up by a reward increment ΔL, whereas, all other links $L_{i,j}^{s,v}$ may be reduced by a punishment increment ΔM.

(Fig.13.2) [Graupe 1997, Graupe and Kordylewski Graupe, 1997]. The values of L-link weights are modified according to:

$$L_{i,j}^{k,m}(t+1) = L_{i,j}^{k,m}(t) + \Delta L : \; L_{i,j}^{k,m} \leq L_{\max} \tag{13.5a}$$

$$L_{i,j}(t+1) = L_{i,j}(t) - \Delta M \tag{13.5b}$$

$$L(0) = 0 \tag{13.5c}$$

where:

$L_{i,j}^{k,m}$: links between winning neuron i in k'th module and winning neuron j in m'th module (which may also be the m'th output module).

ΔL, ΔM: reward/punishment increment values (predetermined fixed values). It is sometimes desirable to set ΔM (either for all LAMSTAR decisions or only when the decision is correct) as:

$$\Delta M = 0 \tag{13.6}$$

$L_{\max}$: maximal links value (not generally necessary, especially when update via forgetting is performed).

The link weights thus serve as address correlations [Graupe and Lynn, 1970] to evaluate traffic rates between neurons [Graupe, 1997, Minsky, 1980]. See Fig. 13.1. The L link weights above thus serve to guide the storage process and to speed it up in problems involving very many subwords (patterns) and huge memory in each such pattern. They also serves to exclude patterns that totally overlap, such that one (or more) of them are redundant and need be omitted. In many applications, the only link weights considered (and updated) are those between the SOM storage layers (modules) and the output layers (as in Fig. 13.2), while link-weights between the various SOM input-storage layers (namely, **internal link-weights**) are **not considered or updated**, unless they are required for decisions related to Sec. 13.2.6 below.

13.2.5. *Determination of winning decision via link weights*

The diagnosis/decision at the output SOM modules is found by analyzing correlation links L between diagnosis/decision neurons in the output SOM modules and the **winning neurons** in all input SOM modules selected and accepted by the process outlined in Sec. 13.2.4. Furthermore, all **L-weight values are set** (updated) as discussed in Sec. 13.2.4 above (Eqs. (13.6), (13.7a) and (13.7b)).

The winning neuron (diagnosis/decision) from the output SOM module is a neuron with the highest cumulative value of links L connecting to the selected (winning) input neurons in the input modules. The diagnosis/detection formula for output SOM module (i) is given by:

$$\sum_{kw}^{M} L_{kw}^{i,n} \geq \sum_{kw}^{M} L_{kw}^{i,j} \qquad \forall\, k, j, n; \quad i \neq n \tag{13.7}$$

where:

i: i'th output module.

n: winning neuron in the i'th output module

kw: winning neuron in the k'th input module.

M: number of input modules.

$L_{kw}^{i,j}$: link weight between winning neuron in input module k and neuron j in i'th output module.

Link weights may be either positive or negative. They are preferably **initiated** at a small random value close to zero, though initialization of all weights at zero (or at some other fixed value) poses no difficulty. If two or more weights are **equal** then a certain decision must be pre-programmed to be given a priority.

13.2.6. N_j *weights (not implemented in most applications)*

The N_j weights of Fig. 13.1 [Graupe and Kordyleski, 1998] are updated by the amount of traffic to a given neuron at a given input SOM module, namely by the accumulative number of subwords stored at a given neuron (subject to adjustments due to forgetting as in Sec. 13.3 below), as determined by Eq. (13.8):

$$\|\underline{x}_i - \underline{w}_{i,m}\| = \min \|\underline{x}_i - \underline{w}_{i,k}\|, \qquad \forall\, k \in \langle l, l+p\rangle; \quad l \sim \{N_{i,j}\} \tag{13.8}$$

where

m: is the winning unit in i'th SOM module (WTA),

$(N_{i,j})$: denoting of the weights to determine the neighborhood of top priority neurons in SOM module i, for the purpose of storage search. In most applications, k covers all neurons in a module and both N_{ij} and l are disregarded, as in Fig. 13.2.

l: denoting the first neuron to be scanned (determined by weights $N_{i,j}$);

$\sim$ denoting proportionality.

The N_j weights of Fig. 13.1 above are only used in huge retrieval/browsing problems. They are initialized at some small random non-zero value (selected from a uniforms distribution) and increase linearly each time the appropriate neuron is chosen as winner.

13.2.7. *Initialization and local minima*

In contrast to most other networks, the LAMSTAR neural network is not sensitive to initialization and will not converge to local minima. All link weights should be initialized with the same constant value, preferably zero. However initialization of the storage weights ω_{ij} of Sec. 13.2.2 and of N_j of Sec. 13.2.6 should be at random (very) low values.

Again, in contrast to most other neural networks, the LAMSTAR will not converge to a local minimum, due to its link -weight punishment/reward structure since punishments will continue at local minima.

13.3. Forgetting Feature

Forgetting is introduced in by a forgetting factor $F(k)$; such that:

$$L(k+1) = L(k) - F\{k\}, L(k) > 0, \quad \forall\, k \tag{13.9}$$

For any link weight L, where k denotes the k'th input word considered and where $F(k)$ is a small increment that varies over time (over k).

In certain realizations of the LAMSTAR, the forgetting adjustment is set as:

$$F(k) = 0 \quad \text{over successive } p-1 \text{ input words considered;} \tag{13.10-a}$$

but

$$F(k) = bL \quad \text{per each } p\text{'th input word} \tag{13.10-b}$$

where L is any link weight and

$$b < 1 \tag{13.10-c}$$

say, $b = 0.5$.

Furthermore, in preferred realizations $L_{\max}$ is unbounded, except for reductions due to forgetting.

Noting the forgetting formula of Eqs. (13.9) and (13.10), link weights $L_{i,j}$ decay over time. Hence, if not chosen successfully, the appropriate $L_{i,j}$ will drop towards zero. Therefore, correlation links L which do not participate in successful diagnosis/decision over time, or lead to an incorrect diagnosis/decision are gradually forgotten. The forgetting feature allows the network to rapidly retrieve very recent information. Since the value of these links decreases only gradually and does not drop immediately to zero, the network can re-retrieve information associated with those links. The forgetting feature of the LAMSTAR network helps to avoid the need to consider a very large number of links, thus contributing to the network efficiency. At the forgetting feature requires storage of link weights and numbering of input words. Hence, in the simplest application of forgetting, old link weights are forgotten (subtracted from their current value) after, say every M input words. The forgetting can be applied gradually rather than stepwise as in Eqs. (13.5) above.

A stepwise Forgetting algorithm can be implemented such that all weights and decisions must have an index number $(k, k = 1, 2, 3, \ldots)$ starting from the very first entry. Also, then one must remember the weights as they are every M (say, $M = 20$) input words. Consequently, one updates ALL weights every $M = 20$ input words by subtracting from EACH weight its stored value to be forgotten.

For example, at input word $k = 100$ one subtracts the weights as of Input Word $k = 20$ (or alternatively X%, say, 50% thereof) from the corresponding weights at input word $k = 100$ and thus one KEEPS only the weights of the last 80 input words. Updating of weights is otherwise still done as before and so is the advancement of k. Again, at input word $k = 120$ one subtracts the weights as of input word $k = 40$ to keep the weights for an input-words interval of duration of, say, $P = 80$, and so on.

Therefore, at $k = 121$ the weights (after the subtraction above) cover experience relating to a period of 81 input words. At $k = 122$, they cover a stored-weights experience over 82 input words ..., at $k = 139$ they covers a period of 99 input words, at $k = 140$ they cover 120–20 input words, since now one subtracted the weights of $k = 40$, etc. Hence, weights cover always a period of no more than 99 input words and no less than 80 input words. Weights must then be stored only every $M = 20$ input words, not per every input word. Note that the $M = 20$ and $P = 80$ input words mentioned are arbitrary. When one wishes to keep data over longer periods, one may set M and P to other values as desired.

Simple applications of the LAMSTAR neural network do not always require the implementation of the forgetting feature. If in doubt about using the forgetting property, it may be advisable to compare performance "***with forgetting***" against "***without forgetting***" (when continuing the training throughout the testing period).

13.4. Training vs. Operational Runs

There is no reason to stop training as the first n sets (input words) of data are only to establish initial weights for the testing set of input words (which are, indeed, normal run situations), which, in LAMSTAR, we can still continue training set by set (input-word by input-word). Thus, the NETWORK continues adapting itself during testing and regular operational runs. The network's performance benefits significantly from continued training while the network does not slow down and no additional complexity is involved. In fact, this does slightly simplify the network's design.

13.4.1. *INPUT WORD for training and for information retrieval*

In applications such as medical diagnosis, the LAMSTAR system is trained by entering the symptoms/diagnosis pairs (or diagnosis/medication pairs). The *training* input word **X** is then of the following form:

$$\underline{X} = [\underline{x}_1^T, \underline{x}_2^T, \ldots, \underline{x}_n^T, \underline{d}_1^T, \ldots, \underline{d}_k^T]^T \tag{13.11}$$

where $\boldsymbol{x}_i$ are input subwords and $\boldsymbol{d}_i$ are subwords representing past outputs of the network (diagnosis/decision). Note also that one or more SOM modules may serve as output modules to output the LAMSTAR's decisions/diagnoses.

The input word of Eqs. (13.2) and (13.11) is set to be a set of coded subword (Sec. 13.1), comprising of coded vector-subwords ($\boldsymbol{x}_i$) that relate to various categories (input dimensions). Also, each SOM module of the LAMSTAR network corresponds to one of the categories of $\underline{x}_i$ such that the number of SOM modules equals the number of subvectors (subwords) $\boldsymbol{x}_n$ and $\boldsymbol{d}$ in $\boldsymbol{X}$ defined by Eq. (13.11).

13.5. Advanced Data Analysis Capabilities

Since all information in the LAMSTAR network is encoded in the correlation links, the LAMSTAR can be utilized as a data analysis tool. In this case the system provides analysis of input data such as evaluating the importance of input subwords, the strengths of correlation between categories, or the strengths of correlation of between individual neurons.

The system's analysis of the input data involves two phases:

(1) training of the system (as outlined in Sec. 13.4)
(2) analysis of the values of correlation links as discussed below.

Since the correlation links connecting clusters (patterns) among categories are modified (increased/decreased) in the training phase, it is possible to single out the links with the highest values. Therefore, the clusters connected by the links with the highest values determine the trends in the input data. In contrast to using data averaging methods, isolated cases of the input data will not affect the LAMSTAR results, noting its forgetting feature. Furthermore, the LAMSTAR structure makes it very robust to missing input subwords.

After the training phase is completed, the LAMSTAR system finds the highest correlation links (link weights) and reports messages associated with the clusters in SOM modules connected by these links. The links can be chosen by two methods: (1) links with value exceeding a pre-defined threshold, (2) a pre-defined number of links with the highest value.

13.5.1. *Feature extraction and reduction in the LAMSTAR NN*

Features can be extracted and reduced in the LAMSTAR network according to the derivations leading to the properties of certain elements of the LAMSTAR network as follows:

Definition I: A feature can be extracted by the matrix $A(i, j)$ where i denotes a winning neuron in SOM storage module j. All winning entries are 1 while the rest are 0. Furthermore, $A(i, j)$ can be reduced via considering properties (b) to (e) below.

(a) The ***most (least) significant subword*** (winning memory neuron) {i} ***over all SOM modules*** (i.e., over the whole NN) ***with respect to a given output decision*** $\{dk\}$ ***and over all input words***, denoted as $[i^*, s^*/dk]$, is given by:

$$[i^*, s^*/dk] : L(i, s/dk) \geq L(j, p/dk) \text{ for any winning neuron } \{j\} \text{ in any module } \{p\} \quad (13.12)$$

where p is not equal to s, $L(j, p/dk)$ denoting the link weight between the j'th (winning) neuron in layer p and the winning output-layer neuron dk. Note that for determining the least significant neuron, the inequality as above is reversed.

(b) The ***most (least) significant SOM module*** $\{s^{**}\}$ ***per a given winning output decision*** $\{dk\}$ ***over all input words***, is given by:

$$s^{**}(dk) : \sum_i(\{L(i, s/dk)\} \geqq \sum_j(\{L(j, p/dk)\} \quad \text{for any module } p \tag{13.13}$$

Note that for determining the least significant module, the inequality above is reversed.

(c) The neuron $\{i^{**}(dk)\}$ that is ***most (least) significant in a particular SOM module (s) per a given output decision*** (dk), ***over all input words per a given class of problems***, is given by $i^*(s, dk)$ such that:

$$L(i, s/dk) \geqq L(j, s/dk) \quad \text{for any neuron } (j) \text{ in same module } (s)\,. \tag{13.14}$$

Note that for determining the least significant neuron in module (s), the inequality above is reversed.

(d) ***Redundancy via Internal Links:*** If the link weights $L(p, a/q, b)$ from any neuron $\{p\}$ in layer $\{a\}$ to some neuron $\{q\}$ in layer $\{b\}$ is very high, WHILE it is (near) zero to EVERY OTHER neuron in layer $\{b\}$, we denote the neuron $\{q\}$ in layer $\{b\}$ as $q(p)$. Now, IF this holds for ALL neurons $\{p\}$ in layer $\{a\}$ which were ever selected (declared winners) , THEN layer $\{b\}$ is REDUNDANT, as long as the number of neurons $\{p\}$ is larger or equal to the number of $\{q(p)\}$, AND layer $\{b\}$ should be removed.

Definition II: If the number of $\{q(p)\}$ neurons is less than the number of $\{p\}$ neurons, then layer $\{b\}$ is called an ***INFERIOR LAYER*** to $\{a\}$.

Also see Property (i) below on redundancy determination via correlation-layers.

(e) ***Zero-Information Redundancy:*** If only one neuron is ALWAYS the winner in layer (k), regardless of the output decision, then the layer contains no information and is redundant.

The above definitions and properties can serve to reduce number of features or memories by considering only a reduced number of most-significant modules or memories or by eliminating the least significant ones.

13.6. Correlation, Interpolation, Extrapolation and Innovation-Detection

13.6.1. *Correlation feature*

Consider the (m) most significant layers (modules) with respect to output decision (dk) and the (n) most significant neurons in each of these (m) layers, with respect to the same output decision. (Example: Let $m = n = 4$). We comment that correlation between subwords can also be accommodated in the network by assigning a specific input subword of that correlation, this subword being formed by pre-processing.

(f) ***Correlation-Layer Set-Up Rule:*** Establish additional SOM layers denoted as CORRELATION-LAYERS $\lambda(p/q, dk)$, such that the number of these additional correlation-layers is:

$$\sum_{i=1}^{m-1} \text{ per output decision } dk \tag{13.15}$$

(Example: The correlation-layers for the case of $n = m = 4$ are: $\lambda(1/2, dk)$; $\lambda(1/3, dk)$; $\lambda(1/4, dk)$; $\lambda(2/3, dk)$; $\lambda(2/4, dk)$; $\lambda(3/4, dk)$.)

Subsequently, WHENEVER neurons $N(i,p)$ and $N(j,q)$ are simultaneously (namely, for the same given input word) winners at layers (p) and (q) respectively, and both these neurons also belong to the subset of 'most significant' neurons in 'most significant' layers (such that p and q are 'most significant' layers), THEN we declare a neuron $N(i,p/j,q)$ in Correlation-Layer $\lambda(p/q, dk)$ to be the winning neuron in that correlation-layer and we reward/punish its output link-weight $L(i,p/j,q - dk)$ as need be for any winning neuron in any other input SOM layer.

(Example: The neurons in correlation-layer $\lambda(p/q)$ are: $N(1,p/1,q)$; $N(1,p/2,q)$; $N(1,p/3,q)$; $N(1,p/4,q)$, $N(2,p/1,q)$; $\ldots N(2,p/4,q)$; $N(3,p/1,q)$; $\ldots N(4,p/1,q)$; $\ldots N(4,p/4,q)$, to total mxm neurons in the correlation-layer).

Any winning neuron in a correlation layer is treated and weighted as any winning neuron in another (input-SOM) layer as far as its weights to any output layer neuron are concerned and updated. Obviously, a winning neuron (per a given input word), if any, in a correlation layer p/q is a neuron $N(i,p/j,q)$ in that layer where both neuron $N(i,p)$ in input layer (p) and neuron $N(j,q)$ in layer (q) were winners for the given input word.

(g) ***Interpolation/Extrapolation via Internal Link:*** For a given input word that relates to output decision dk, if no input subword exists that relates to layer (p), then the neuron $N(i,p)$ which has the highest summed-correlation link (internal link weights) with winning neurons (for the same input word) in other layers v, will be considered the interpolation/extrapolation neuron in layer p for that input word. However, no rewards/punishments will be applied to that neuron while it is an interpolation/extrapolation neuron.

(h) ***Interpolation/Extrapolation via Correlation Layers:*** Let p be a 'most significant' layer and let i be a 'most significant neuron with respect to output decision dk in layer p, where no input subword exists in a given input word relating to layer p. Thus, neuron $N(i,p)$ is considered as the interpolation/extrapolation neuron for layer p if it satisfies:

$$\sum_q \{L(i,p/w,q - dk)\} \geqq \sum_q \{L(v,p/w,q - dk)\} \tag{13.16}$$

where v are different from i and where $L(i,p/j,q - dk)$ denote link weights from correlation-layer $\lambda(p/q)$. Note that in every layer q there is only one winning neuron for the given input word, denoted as $N(w,q)$, whichever w may be at any q'th, layer.

(Example: Let $p = 3$. Thus consider correlation-layers $\lambda(1/3, dk)$; $\lambda(2/3, d\kappa)$; $\lambda(3/4, dk)$ such that: $q = 1, 2, 4$.)

(i) ***Redundancy via Correlation-Layers:*** Let p be a 'most significant' layer and let i be a 'most significant' neuron in that layer. Layer p is redundant if for all input words there is there is another 'most significant' layer q such that, for any output decision and for any neuron $N(i,p)$, only one correlation neuron $i,p/j,q$ (i.e., for only one j per each such i,p) has non-zero output-link weights to any output decision dk, such that every neuron $N(j,p)$ is always associated with only one neuron $N(j,p)$ in some layer p.

(Example: Neuron $N(1,p)$ is always associated with neuron $N(3,q)$ and never with $N(1,q)$ or $N(2,q)$ or $N(4,q)$, while neuron $N(2,p)$ is always associated with $N(4,q)$ and never with other neurons in layer q).

Also, see property (d) above.

13.6.2. *Innovation detection in the LAMSTAR NN*

(j) If link-weights from a given input SOM layer to the output layer output change considerably and repeatedly (beyond a threshold level) within a certain time interval (a certain specified number of successive input words that are being applied), relatively to link weights from other input SOM layers, then ***innovation*** is detected with respect to that input layer (category).

(k) ***Innovation*** is also detected if weights between neurons from one input SOM layer to another input SOM layer similarly change.

13.7. Concluding Comments and Discussion of Applicability

The LAMSTAR neural network utilizes the basic features of many other neural network, and adopts Kohonen's SOM modules [Kohonen, 1977, 1984] with their associative-memory — based setting of storage weights (w_{ij} in this Chapter) and its WTA (Winner-Take-All) feature, it differs in its neuronal structure in that every neuron has not only storage weights w_{ij} (see Chapter 8 above), but also the link weights L_{ij}. This feature directly follows Hebb's Law [Hebb, 1949] and its relation to Pavlov's Dog experiment, as discussed in Sec. 3.1. It also follows Minsky's k-lines model [Minsky, 1980] and Kant's emphasis on the essential role Verbindungen in "understanding", which obviously relates to the ability to decide. Hence, not only does LAMSTAR deal with two kinds of neuronal weights (for storage and for linkage to other layers), but in the LAMSTAR, the link weights are the ones that count for decision purposes. The storage weights form "atoms of memory" in the Kantian sense [Ewing, 1938]. The LAMSTAR's decisions are solely based on these link weights — see Sec. 13.2 below.

The LAMSTAR, like all other neural networks, attempts to provide a representation of the problem it must solve (Rosenblatt, 1961). This representation, regarding the networks decision, can be formulated in terms of a nonlinear mapping **L** of the weights between the inputs (input vector) and the outputs, that is arranged in a matrix form. Therefore, **L** is a nonlinear mapping function whose entries are the weights between inputs an the outputs, which map the inputs to the output decision. Considering the Back-Propagation (BP) network, the weights in each layer are the columns of **L**. The same holds for the link weights L_{ij} of **L** to a winning output decision in the LAMSTAR network. Obviously, in both BP and LAMSTAR, **L** is not a square matrix-like function, nor are all its columns of same length. However, in BP, **L** has ***many entries*** (weights) in ***each*** column per any output decision. In contrast, in the LAMSTAR, each column of **L** has ***only one non-zero entry***. This accounts both for the ***speed*** and the ***transparency*** of LAMSTAR. There weights in BP do not yield direct information on what their values mean. In the LAMSTAR, the link weights directly indicate the significance of a given feature and of a particular subword relative to the particular decision, as indicated in Sec. 13.5 below. The basic LAMSTAR algorithm require the computation of only Eqs. (13.5) and (13.7) per iteration. These involve only addition/subtraction and thresholding operations while no multiplication is involved, to further contribute to the LAMSTAR's computational speed.

The LAMSTAR network facilitates a multidimensional analysis of input variables to assign, for example, different weights (importance) to the items of data, find correlation among input variables, or perform identification, recognition and clustering of patterns. Being a neural network, the LAMSTAR can do all this without re-programming for each diagnostic problem.

The decisions of the LAMSTAR neural network are based on many categories of data, where often some categories are fuzzy while some are exact, and often categories are missing (incomplete data sets). As mentioned in Sec. 13.1 above, the LAMSTAR network can be trained with incomplete data or category sets. Therefore, due to its features, the LAMSTAR neural network is a very effective tool in just such situations. As an input, the system accepts data defined by the user, such as, system state, system parameters, or very specific data as it is shown in the application examples presented below. Then, the system builds a model (based on data from past experience and training) and searches the stored knowledge to find the best approximation/description to the features/parameters given as input data. The input data could be automatically sent through an interface to the LAMSTAR's input from sensors in the system to be diagnosed, say, an aircraft into which the network is built in.

The LAMSTAR system can be utilized as:

- Computer-based medical diagnosis system [Kordylewski and Graupe, 2001, Nigam and Graupe, 2004, Muralidharan and Rousche, 2005].
- Tool for financial evaluations.

— Tool for industrial maintenance and fault diagnosis (on same lines as applications to medical diagnosis).
— Tool for data mining [Carino et al., 2005].
— Tool for browsing and information retrieval .
— Tool for data analysis, classification, browsing, and prediction [Sivaramakrishnan and Graupe, 2004].
— Tool for image detection and recognition [Girado et al., 2004].
— Teaching aid.
— Tool for analyzing surveys and questionnaires on diverse items.

All these applications can employ many of the other neural networks that we discussed. However, the LAMSTAR has certain advantages, such as insensitivity to initialization, the avoidance of local minima, its forgetting capability (this can often be implemented in other networks), its transparency (the link weights carry clear information as to the link weights on relative importance of certain inputs, on their correlation with other inputs, on innovation detection capability and on redundancy of data — see Secs. 13.5 and 13.6 above). The latter allow downloading data without prior determination of its significance and letting the network decide for itself , via the link weights to the outputs. The LAMSTAR, in contrast to many other networks, can work uninterrupted if certain sets of data (input-words) are incomplete (missing subwords) without requiring any new training or algorithmic changes. Similarly, input subwords can be added during the network's operation without reprogramming while taking advantage of its forgetting feature. Furthermore, the LAMSTAR is very fast, especially in comparison to back-propagation or to statistical networks, without sacrificing performance and it always learns during regular runs.

Appendix 13.A provides details of the LAMSTAR algorithm for the Character Recognition problem that was also the subject of Appendices to Chapters 5, 6, 7, 8, 9, 11 and 12. Examples of applications to medical decision and diagnosis problems are given in Appendix 13.B below.

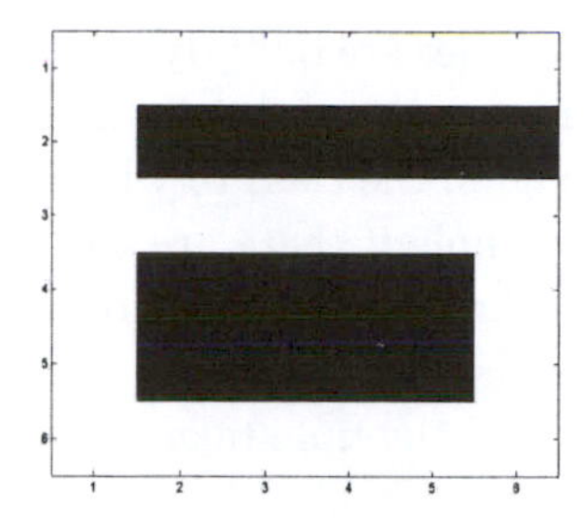

Fig. 13.A. 1: Example of a training pattern ('6').

1	1	1	1	1	1
1	0	0	0	0	0
1	1	1	1	1	1
1	0	0	0	0	1
1	0	0	0	0	1
1	1	1	1	1	1

Fig. 13.A.2: Unipolar Representation of '6'.

13.A. LAMSTAR Network Case Study*: Character Recognition

13.A.1. *Introduction*

This case study focuses on recognizing characters '6', '7', 'X' and "rest of the world" patterns namely, patterns not belonging to the set '6', '7', 'X'). The characters in the training and testing set are represented as unipolar inputs '1' and '0' in a $6 * 6$ grid. An example of a character is as follows:

13.A.2. *Design of the network*

The LAMSTAR network has the following components:

(a) *INPUT WORD AND ITS SUBWORDS:*

The input word (in this case, the character) is divided into a number of subwords. Each subword represents an attribute of the input word. The subword division in the character recognition problem was done by considering every row and every column as a subword hence resulting in a total of 12 subwords for a given character.

(b) *SOM MODULES FOR STORING INPUT SUBWORDS:*

For every subword there is an associated Self Organizing Map (SOM) module with neurons that are designed to function as Kohonen 'Winner Take All' neurons where the winning neuron has an output of 1 while all other neurons in that SOM module have a zero output.

*Computed by Vasanth Arunachalam, ECE Dept., University of Illinois, Chicago, 2005.

In this project, the SOM modules are built dynamically in the sense that instead of setting the number of neurons at some fixed value arbitrarily, the network was built to have neurons depending on the class to which a given input to a particular subword might belong. For example if there are two subwords that have all their pixels as '1's, then these would fire the same neuron in their SOM layer and hence all they need is 1 neuron in the place of 2 neurons. This way the network is designed with lesser number of neurons and the time taken to fire a particular neuron at the classification stage is reduced considerably.

(c) *OUTPUT (DECISION) LAYER:*

The present output layer is designed to have two layers, which have the following neuron firing patterns:

Table 13.A.1: Firing order of the output neurons.

Pattern	Output Neuron 1	Output Neuron 2
'6'	Not fired	Not fired
'7'	Not fired	Fired
'X'	Fired	Not fired
'Rest of the World'	Fired	Fired

The link-weights from the input SOM modules to the output decision layer are adjusted during training on a reward/punishment principle. Furthermore, they continue being trained during normal operational runs. Specifically, if the output of the particular output neuron is what is desired, then the link weights to that neuron is rewarded by increasing it by a non-zero increment, while punishing it by a small non-zero number if the output is not what is desired.

Note: The same can be done (correlation weights) between the winning neurons of the different SOM modules but has not been adopted here due to the complexities involved in implementing the same for a generic character recognition problem.

The design of the network is illustrated in Fig. 13.A.3.

13.A.3. *Fundamental principles*

Fundamental principles used in dynamic SOM layer design

As explained earlier the number of neurons in every SOM module is not fixed. The network is designed to grow dynamically. At the beginning there are no neurons in any of the modules. So when the training character is sent to the network, the first neuron in every subword is built. Its output is made 1 by adjusting the weights based on the 'Winner Take All' principle. When the second training pattern is input to the system, this is given as input to the first neuron and if the output is close to 1 (with a tolerance value of 0.05), then the same neuron is fired and another neuron

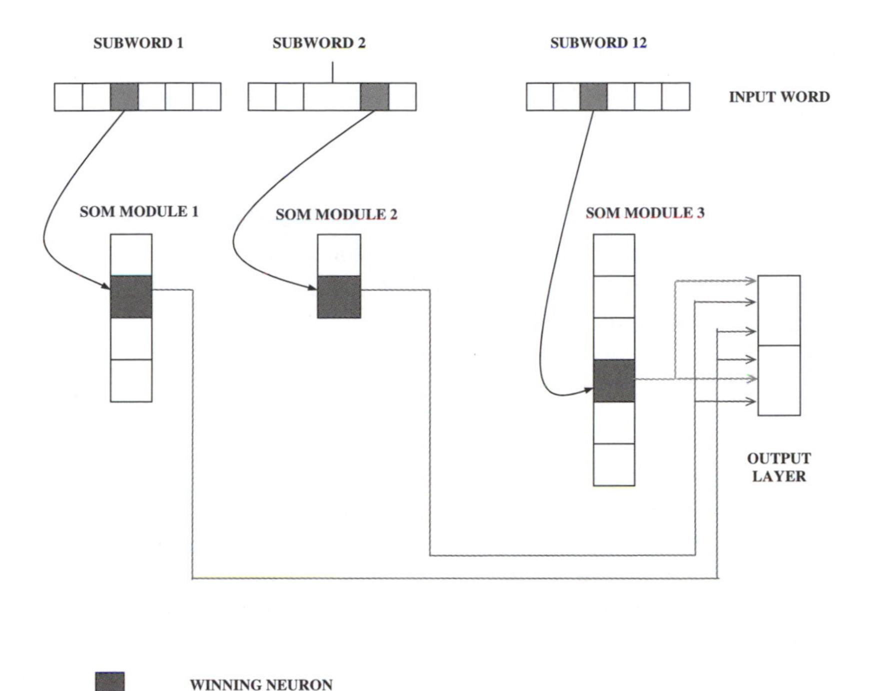

Fig. 13.A.3. Design of the LAMSTAR neural network for character recognition. Number of SOM modules in the network is 12. The neurons (Kohonen) are designed to build dynamically which enables an adaptive design of the network. Number of neurons in the output layer is 2. There are 12 subwords for every character input to the network. Green denotes the winning neuron in every SOM module for the respective shaded subword pixel. Reward/Punishment principle is used for the output weights.

is not built. The second neuron is built only when a distinct subword appears at the input of all the previously built neuron resulting in their output not being sufficiently close to 1 so as to declare any of them a winning neuron.

It has been observed that there has been a significant reduction in the number of neurons required in every SOM modules.

Winner Take All principle

The SOM modules are designed to be Kohonen layer neurons, which act in accordance to the 'Winner Take All' Principle. This layer is a competitive layer wherein the Eucledian distance between the weights at every Kohonen layer and the input pattern is measured and the neuron that has the least distance if declared to be the winner. This Kohonen neuron best represents the input and hence its output is made equal to 1 whereas all other neuron outputs are forced to go to 0. This principle is called the 'Winner Take All' principle. During training the weights corresponding to the winning neuron is adjusted such that it closely resembles the input pattern while all other neurons move away from the input pattern.

13.A.4. *Training algorithm*

The training of the LAMSTAR network if performed as follows:

(i) **Subword Formation:**
The input patterns are to be divided into subwords before training/testing the LAMSTAR network. In order to perform this, the every row of the input 6*6 character is read to make 6 subwords followed by every column to make another 6 subwords resulting in a total of 12 subwords.

(ii) **Input Normalization:**
Each subwords of every input pattern is normalized as follows:

$$xi' = x_i \Big/ \sqrt{\Sigma x_j^2}$$

where, x — subword of an input pattern. During the process, those subwords, which are all zeros, are identified and their normalized values are manually set to zero.

(iii) **Rest of the world Patterns:**
The network is also trained with the rest of the world patterns 'C', 'I' and '||'. This is done by taking the average of these patterns and including the average as one of the training patterns.

(iv) **Dynamic Neuron formation in the SOM modules:**
The first neuron in all the SOM modules are constructed as Kohonen neurons as follows:

- As the first pattern is input to the system, one neuron is built with 6 inputs and random weights to start with initially and they are also normalized just like the input subwords. Then the weights are adjusted such that the output of this neuron is made equal to 1 (with a tolerance of 10^{-5} according to the formula:

$$w(n+1) = w(n) + \alpha^*(x - w(n))$$

where,
α — learning constant = 0.8
w — weight at the input of the neuron
x — subword

$$z = w^*x$$

where, z — output of the neuron (in the case of the first neuron it is made equal to 1).

- When the subwords of the subsequent patterns is input to the respective modules, the output at any of the previously built neuron is checked to see if it is close to 1 (with a tolerance of 0.05). If one of the neurons satisfies the condition, then this is declared as the winning neuron, i.e., a neuron whose weights closely resemble the input pattern. Else another neuron is built with new sets of weights that are normalized and adjusted as above to resemble the input subword.

- During this process, if there is a subword with all zeros then this will not contribute to a change in the output and hence the output is made to zero and the process of finding a winning neuron is bypassed for such a case.

(v) **Desired neuron firing pattern:**
The output neuron firing pattern for each character in the training set has been established as given in Table 1.

(vi) **Link weights:**
Link weights are defined as the weights that come from the winning neuron at every module to the 2 output neurons. If in the desired firing, a neuron is to be fired, then its corresponding link weights are rewarded by adding a small positive value of 0.05 every iteration for 20 iterations. On the other hand, if a neuron should not be fired then its link weights are reduced 20 times by 0.05. This will result in the summed link weights at the output layer being a positive value indicating a fired neuron if the neuron has to be fired for the pattern and high negative value if it should not be fired.

(vii) The weights at the SOM neuron modules and the link weights are stored.

13.A.4.1. *Training set*

The LAMSTAR network is trained to detect the characters ‘6’, ‘7’, ‘X’ and ‘rest of the world’ characters. The training set consists of 16 training patterns 5 each for ‘6’, ‘7’ and ‘X’ and one average of the ‘rest of the world’ characters.

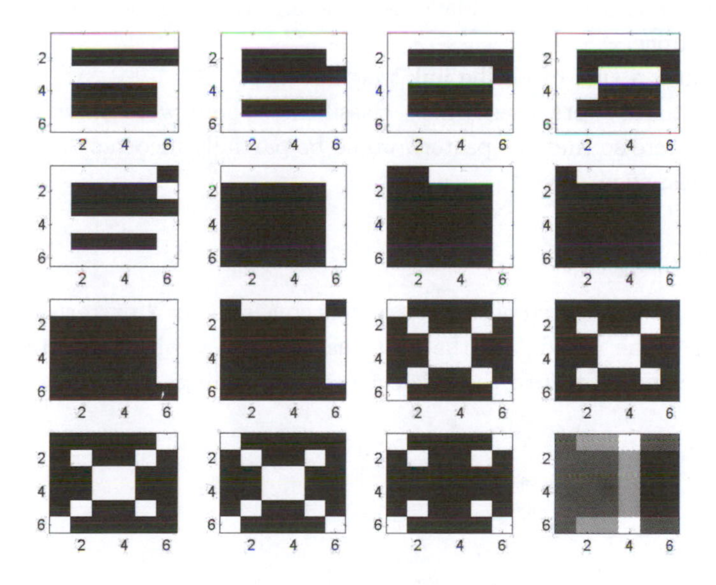

Fig. 13.A.4. Training Pattern Set for recognizing characters ‘6’, ‘7’, ‘X’ and ‘mean of rest of world’ patterns ‘C’, ‘I’, ‘||’.

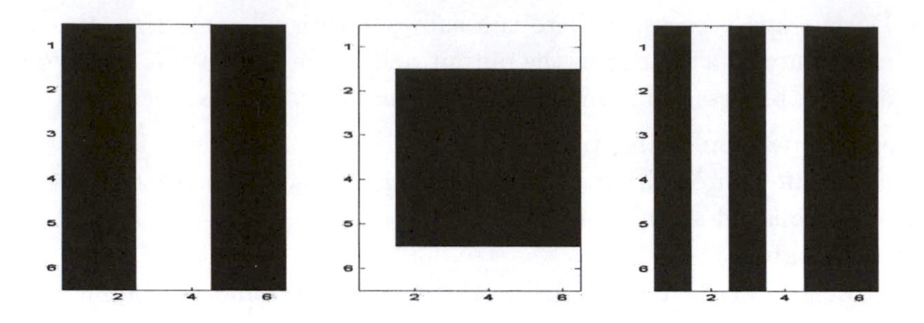

Fig. 13.A.5. 'Rest of the world patterns 'I', 'C' and '||'.

13.A.4.2. *'Rest of the world' patterns*

The rest of the world patterns used to train the network are as follows:

13.A.5. *Testing procedure*

The LAMSTAR network was tested with 8 patterns as follows:

- The patterns are processed to get 12 subwords as before. Normalization is done for the subwords as explained in the training.
- The stored weights are loaded
- The subwords are propagated through the network and the neuron with the maximum output at the Kohonen layer is found and their link weights are sent to the output neurons.
- The output is a sum of all the link weights.
- All the patterns were successfully classified. There were subwords that were completely zero so that the pattern would be partially incorrect. Even these were correctly classified.

13.A.5.1. *Test pattern set*

The network was tested with 8 characters consisting of 2 pattern each of '6', '7', 'X' and rest of the world. All the patterns are noisy, either distorted or a whole row/column removed to test the efficiency of the training. The following is the test pattern set.

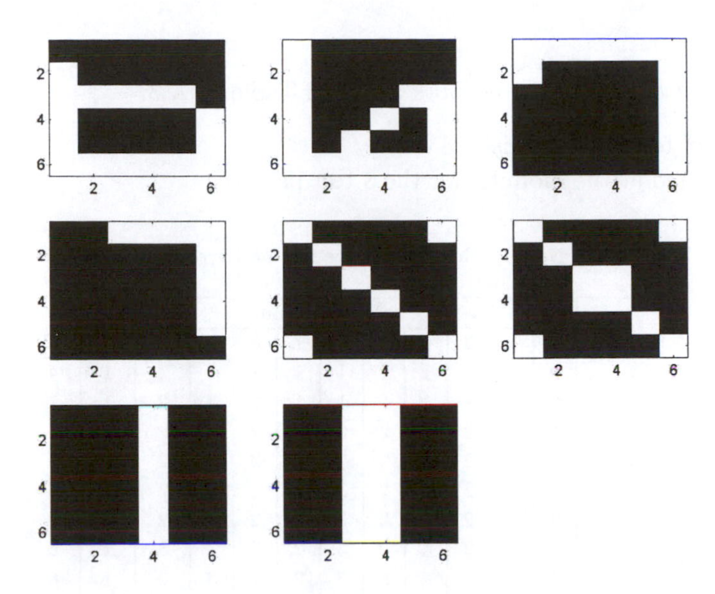

Fig. 13.A.6. Test pattern set consisting of 2 patterns each for '2', '7', 'X' and 'rest of the world'.

13.A.6. *Results and their analysis*

13.A.6.1. *Training results*

The results obtained after training the network are presented in Table 13.A.2:

- Number of training patterns = 16
- Training efficiency = 100%
- Number of SOM modules = 12
- The number of neurons in the 12 SOM modules after dynamic neuron formation in are:

Table 13.A.2. Number of neurons in the SOM modules.

SOM Module Number	Number of neurons
1	3
2	2
3	2
4	4
5	2
6	4
7	3
8	3
9	3
10	3
11	3
12	7

13.A.6.2. *Test results*

The result of testing the network are as in Table 13.A.3:

- Number of testing patterns = 8
- Neurons fired at the modules for the 8 test patterns:

Table 13.A.3: Neurons fired during the testing for respective patterns.

Pattern	Module Number											
	1	2	3	4	5	6	7	8	9	10	11	12
6	0	0	0	1	1	1	1	1	1	1	1	4
6	0	0	1	1	1	1	1	2	3	2	1	1
7	1	1	2	4	2	2	1	2	3	2	1	5
7	1	2	2	4	2	2	1	2	3	2	1	5
X	2	2	2	4	2	3	2	2	3	2	1	6
X	2	2	2	4	2	3	2	2	3	2	1	6
|	2	2	2	4	2	3	2	2	3	3	1	6
||	2	2	2	4	2	3	2	2	3	3	1	6

The firing pattern of the output neurons for the test set is given in Table 13.A.4:

Table 13.A.4: Firing pattern for the test characters.

Test Pattern	Neuron 1	Neuron 2
6 (with bit error)	−25.49 (Not fired)	−25.49 (Not fired)
6 (with bit error)	−20.94 (Not fired)	−20.94 (Not fired)
7 (with bit error)	−29.99 (Not fired)	15.99 (Fired)
7 (with bit error)	−24.89 (Not fired)	18.36 (Fired)
X (with bit error)	9.99 (Fired)	−7.99 (Not fired)
X (with bit error)	9.99 (Fired)	−7.99 (Not fired)
| (with bit error)	0.98 (Fired)	0.98 (Fired)
|| (with bit error)	1.92 (Fired)	1.92 (Fired)

- Efficiency: 100%.

13.A.7. *Summary and concluding observations*

Summary:

- Number of training patterns = 16 (5 each of '6', '7', 'X' and 1 mean image of 'rest of the world'
- Number of test patterns = 8 (2 each for '6', '7', 'X' and 'rest of the world' with bit errors)
- Number of SOM modules = 12

- Number of neurons in the output layer = 2
- Number of neurons in the SOM module changes dynamically. Refer table 2 for the number of neurons in each module.
- Efficiency = 100%

Observations:

- The network was much faster than the Back Propagation network for the same character recognition problem.
- By dynamically building the neurons in the SOM modules, the number of computations is largely reduced as the search time to find the winning neuron is reduced to a small number of neurons in many cases.
- Even in the case when neurons are lost (simulated as a case where the output of the neuron is zero i.e., all its inputs are zeros), the recognition efficiency is 100%. This is attributed to the link weights, which takes cares of the above situations.
- The NN learns as it goes even if untrained
- The test patterns where all noisy (even at several bits, yet efficiency was 100%.

13.A.8. *LAMSTAR CODE (MATLAB)*

Main.m

```
clear all
close all

X = train_pattern;
%pause(1)
%close all

n = 12 % Number of subwords
flag = zeros(1,n);

% To make 12 subwords from 1 input
for i = 1:min(size(X)),
   X_r{i} = reshape(X(:,i),6,6);
   for j = 1:n,
   if (j<=6),
       X_in{i}(j,:) = X_r{i}(:,j)';
   else
       X_in{i}(j,:) = X_r{i}(j-6,:);
   end
end

% To check if a subword is all '0's and makes it normalized value equal to zero
% and to normalize all other input subwords
p(1,:) = zeros(1,6);
for k = 1:n,
   for t = 1:6,
      if (X_in{i}(k,t)~= p(1,t)),
          X_norm{i}(k,:) = X_in{i}(k,:)/sqrt(sum(X_in{i}(k,:).^2));
```

```
        else
            X_norm{i}(k,:) = zeros(1,6);
        end
    end
end
end%%%End of for

%%%%%%%%%%%%%%%%%%%%%%%%%%%%%%
% Dynamic Building of neurons
%%%%%%%%%%%%%%%%%%%%%%%%%%%%%%
% Building of the first neuron is done as Kohonen Layer neuron
%(this is for all the subwords in the first input pattern for all SOM modules

i = 1;
ct = 1;
while (i<=n),
 i
 cl = 0;
 for t = 1:6,
    if (X_norm{ct}(i,t)==0),
        cl = cl+1;
    end
 end
 if (cl == 6),
     Z{ct}(i) = 0;
 elseif (flag(i) == 0),
    W{i}(:,ct) = rand(6,1);
    flag(i) = ct;
    W_norm{i}(:,ct) = W{i}(:,ct)/sqrt(sum(W{i}(:,ct).^2));
    Z{ct}(i)= X_norm{ct}(i,:)*W_norm{i};

    alpha =0.8;
    tol = 1e-5;

    while(Z{ct}(i) <= (1-tol)),
      W_norm{i}(:,ct) = W_norm{i}(:,ct) + alpha*(X_norm{ct}(i,:)' -
 W_norm{i}(:,ct));
      Z{ct}(i) =  X_norm{ct}(i,:)*W_norm{i}(:,ct);
    end%%%%End of while
   end%%%%End of if
   r(ct,i) = 1;
   i = i+1;
 end%%%%End of while


 r(ct,:) = 1;
 ct = ct+1;
 while (ct <= min(size(X))),
   for i = 1:n,
      cl = 0;
      for t = 1:6,
         if (X_norm{ct}(i,t)==0),
             cl = cl+1;
         end
      end
```

```
  if (cl == 6),
      Z{ct}(i) = 0;
  else
     i
     r(ct,i) = flag(i);
     r_new=0;
     for k = 1:max(r(ct,i)),
        Z{ct}(i) = X_norm{ct}(i,:)*W_norm{i}(:,k);

        if Z{ct}(i)>=0.95,
           r_new = k;
           flag(i) = r_new;
           r(ct,i) = flag(i);
           break;
        end%%%End of if
     end%%%%%%%End of for

     if (r_new==0),
         flag(i) = flag(i)+1;
         r(ct,i) = flag(i);
         W{i}(:,r(ct,i)) = rand(6,1);
         %flag(i) = r
         W_norm{i}(:,r(ct,i)) = W{i}(:,r(ct,i))/sqrt(sum(W{i}(:,r(ct,i)).^2));
         Z{ct}(i) = X_norm{ct}(i,:)*W_norm{i}(:,r(ct,i));

         alpha =0.8;
         tol = 1e-5;

         while(Z{ct}(i) <= (1-tol)),
           W_norm{i}(:,r(ct,i)) = W_norm{i}(:,r(ct,i)) + alpha*(X_norm{ct}(i,:)' -
  W_norm{i}(:,r(ct,i)));
           Z{ct}(i) =  X_norm{ct}(i,:)*W_norm{i}(:,r(ct,i));
         end%%%End of while
      end%%%End of if
      %r_new
      %disp('Flag')
      %flag(i)
    end%%%%End of if
  end

  ct = ct+1;
end
save W_norm W_norm

for i = 1:5,
    d(i,:) = [0 0];
    d(i+5,:) = [0 1];
    d(i+10,:) = [1 0];
end
d(16,:) = [1 1];
%%%%%%%%%%%%%%%
% Link Weights
%%%%%%%%%%%%%%%
ct = 1;
```

```
    m_r = max(r);
    for i = 1:n,
       L_w{i} = zeros(m_r(i),2);
       end
    ct = 1;

    %%% Link weights and output calculations
    Z_out = zeros(16,2);
    while (ct <= 16),
      ct
      %for mn = 1:2
      L = zeros(12,2);
      %    for count = 1:20,
      for i = 1:n,
        if (r(ct,i)~=0),
           for j = 1:2,
               if (d(ct,j)==0),
                   L_w{i}(r(ct,i),j) = L_w{i}(r(ct,i),j)-0.05*20;
               else
                   L_w{i}(r(ct,i),j) = L_w{i}(r(ct,i),j)+0.05*20;
               end %%End if loop
           end %%% End for loop
           L(i,:) = L_w{i}(r(ct,i),:);
        end %%%End for loop
      end
      %   end %%% End for loop
      Z_out(ct,:) = sum(L);
   ct = ct+1;
   end

save L_w L_w
```

Test.m

```
clear all
X = test_pattern;
load W_norm
load L_w

% To make 12 subwords
for i = 1:min(size(X)),
   i
   X_r{i} = reshape(X(:,i),6,6);
   for j = 1:12,
       if (j<=6),
          X_in{i}(j,:) = X_r{i}(:,j)';
       else
          X_in{i}(j,:) = X_r{i}(j-6,:);
       end
   end

   p(1,:) = zeros(1,6);

   for k = 1:12,
```

```
      for t = 1:6,
         if (X_in{i}(k,t)~= p(1,t)),
             X_norm{i}(k,:) = X_in{i}(k,:)/sqrt(sum(X_in{i}(k,:).^2));
         else
             X_norm{i}(k,:) = zeros(1,6);
         end
      end
   end

   for k = 1:12,
      Z = X_norm{i}(k,:)*W_norm{k};
      if (max(Z) == 0),
          Z_out(k,:) = [0 0];
      else
        index(k) = find(Z == max(Z));
        L(k,:) = L_w{k}(index(k),:);
        Z_out(k,:) = L(k,:)*Z(index(k));
      end
   end
   final_Z = sum(Z_out)
end
```

training_pattern.m

```
function train = train_pattern

x1 = [1 1 1 1 1 1; 1 0 0 0 0 0; 1 1 1 1 1 1; 1 0 0 0 0 1; 1 0 0 0 0 1;
      1 1 1 1 1 1];
x2 = [1 1 1 1 1 1; 1 0 0 0 0 1; 1 0 0 0 0 0; 1 1 1 1 1 1; 1 0 0 0 0 1;
      1 1 1 1 1 1];
x3 = [1 1 1 1 1 1; 1 0 0 0 0 0; 1 1 1 1 1 0; 1 0 0 0 0 1; 1 0 0 0 0 1;
      1 1 1 1 1 1];
x4 = [1 1 1 1 1 1; 1 0 0 0 0 0; 1 0 1 1 1 0; 1 1 0 0 0 1; 1 0 0 0 0 1;
      1 1 1 1 1 1];
x5 = [1 1 1 1 1 0; 1 0 0 0 0 1; 1 0 0 0 0 0; 1 1 1 1 1 1; 1 0 0 0 0 1;
      1 1 1 1 1 1];

x6 = zeros(6,6);
x6(1,:) = 1;
x6(:,6) = 1;

x7 = zeros(6,6);
x7(1,3:6) = 1;
x7(:,6) = 1;

x8 = zeros(6,6);
x8(1,2:6) = 1;
x8(:,6) = 1;

x9 = zeros(6,6);
x9(1,:) = 1;
x9(1:5,6) = 1;

x10 = zeros(6,6);
x10(1,2:5) = 1;
```

```
x10(2:5,6) = 1;

x11 = zeros(6,6);
for i = 1:6,
x11(i,i) = 1;
end
x11(1,6) = 1;
x11(2,5) = 1;
x11(3,4) = 1;
x11(4,3) = 1;
x11(5,2) = 1;
x11(6,1) = 1;

x12 = x11;

x12(1,1) = 0;
x12(6,6) = 0;
x12(1,6) = 0;
x12(6,1) = 0;

x13 = x11;
x13(1,1) = 0;
x13(6,6) = 0;

x14 = x11;
x14(1,6) = 0;
x14(6,1) = 1;

x15 = x11;
x15(3:4,3:4) = 0;

x16 = zeros(6,6);
x16(:,3:4) = 1;

x17 = zeros(6,6);
x17(1,:) = 1;
x17(6,:) = 1;
x17(:,1) = 1;

x18 = zeros(6,6);
x18(:,2) = 1;

x18(:,4) = 1;

x19 = (x16+x17+x18)/3;

xr1 = reshape(x1',1,36);
xr2 = reshape(x2',1,36);
xr3 = reshape(x3',1,36);
xr4 = reshape(x4',1,36);

xr5 = reshape(x5',1,36);
xr6 = reshape(x6',1,36);
xr7 = reshape(x7',1,36);
xr8 = reshape(x8',1,36);
```

```
xr9 = reshape(x9',1,36);
xr10 = reshape(x10',1,36);
xr11 = reshape(x11',1,36);
xr12 = reshape(x12',1,36);

xr13 = reshape(x13',1,36);
xr14 = reshape(x14',1,36);
xr15 = reshape(x15',1,36);

xr19 = reshape(x19',1,36);


xr16 = reshape(x16',1,36);
xr17 = reshape(x17',1,36);
xr18 = reshape(x18',1,36);


train = [xr1' xr2' xr3' xr4' xr5' xr6' xr7' xr8' xr9' xr10' xr11' xr12'
        xr13' xr14' xr15' xr19'];
rest = [xr16' xr17' xr18'];
```

test_pattern.m

```
function t_pat = test_pattern
x1 = [0 0 0 0 0 0; 1 0 0 0 0 0; 1 1 1 1 1 0; 1 0 0 0 0 1; 1 0 0 0 0 1;
      1 1 1 1 1 1];
x2 = zeros(6,6);
x2(:,1) = 1;
x2(3:6,6) = 1;
x2(6,:) = 1;
x2(3,5) = 1;
x2(4,4) = 1;
x2(5,3) = 1;

3 = zeros(6,6);
x3(1,:) = 1;
x3(:,6) = 1;
x3(1:2,1) = 1;

x4 = zeros(6,6);
x4(1,3:6) = 1;
x4(1:5,6) = 1;

x5 = zeros(6,6);
for i = 1:6,
   x5(i,i) = 1;
end
x5(1,6) = 1;
x5(6,1) = 1;

x6 = x5;
x6(3,4) = 1;
x6(4,3) = 1;
```

```
x7 = zeros(6,6);
x7(:,4) = 1;

x8 = zeros(6,6);
x8(:,3:4) = 1;

xr1 = reshape(x1',1,36);
xr2 = reshape(x2',1,36);
xr3 = reshape(x3',1,36);
xr4 = reshape(x4',1,36);
xr5 = reshape(x5',1,36);
xr6 = reshape(x6',1,36);
xr7 = reshape(x7',1,36);
xr8 = reshape(x8',1,36);

t_pat = [xr1' xr2' xr3' xr4' xr5' xr6' xr7' xr8'];
```

13.B. Application to Medical Diagnosis Problems

(a) Application to ESWL Medical Diagnosis Problem

In this application, the LAMSTAR network serves to aid in a typical urological diagnosis problem that is, in fact, a prediction problem [Graupe 1997, Kordylewski et al., 1999]. The network evaluates a patient's condition and provides long term forecasting after removal of renal stones via Extracorporeal Shock Wave Lithotripsy (denoted as ESWL). The ESWL procedure breaks very large renal stones into small pieces that are then naturally removed from the kidney with the urine. Unfortunately, the large kidney stones appear again in 10% to 50% of patients (1–4 years post surgery). It is difficult to predict with reasonable accuracy (more than 50%) if the surgery was a success or a failure, due to the large number of analyzed variables. In this particular example, the input data (denoted as a "word" for each analyzed case, namely, for each patient) are divided into 16 subwords (categories). The length in bytes for each subword in this example varies from 1 to 6 bytes. The subwords describe patient's physical and physiological characteristics, such as patient demographics, stone's chemical composition, stone location, laboratory assays, follow-up, re-treatments, medical therapy, etc.

Table 13.B.1 below compares results for the LAMSTAR network and for a Back-Propagation (BP) neural network [Niederberger et al., 1996], as applied to exactly the same training and test data sets [Kordylewski et al., 1999]. While both networks model the problems with high accuracy, the results show that the LAMSTAR network is over 1000 times faster in this case. The difference in training time is due to the incorporation of an unsupervised learning scheme in the LAMSTAR network, while the BP network training is based on error minimization in a 37-dimensional space (when counting elements of subword vectors) which requires over 1000 iterations.

Both networks were used to perform the Wilks' Lambda test [Morrison, 1996, Wilks, 1938] which serves to determine which input variables are meaningful with

Table 13.B.1. Performance comparison of the LAMSTAR network and the BP network for the renal cancer and the ESWL diagnosis.

	Renal Cancer Diagnosis		ESWL Diagnosis	
	LAMSTAR Network	BP Network Network	LAMSTAR Network	BP Network
Training Time	0.08 sec	65 sec	0.15 sec	177 sec
Test Accuracy	83.15%	89.23%	85.6%	78.79%
Negative Specificity	0.818	0.909	0.53	0.68
Positive Predictive Value	0.95	0.85	1	0.65
Negative Predictive Value	0.714	0.81	0.82	0.86
Positive Specificity	0.95	0.85	1	0.83
Wilks' Test Computation Time	$<$ 15 mins	weeks	$<$ 15 mins	Weeks

Comments:
Positive/Negative Predictive Values — ratio of the positive/negative cases that are correctly diagnosed to the positive/negative cases diagnosed as negative/positive.
Positive/Negative Specificity — he ratio of the positive/negative
cases that are correctly diagnosed to the negative/positive cases that are incorrectly diagnosed as positive/negative.

regard to system performance. In clinical settings, the test is used to determine the importance of specific parameters in order to limit the number of patient's examination procedures.

(b) Application to Renal Cancer Diagnosis Problem

This application illustrates how the LAMSTAR serves to predict if patients will develop a metastatic disease after surgery for removal of renal-cell-tumors. The input variables were grouped into sub-words describing patient's demographics, bone metastases, histologic subtype, tumor characteristics, and tumor stage [Kordylewski et al., 1999]. In this case study we used 232 data sets (patient record), 100 sets for training and 132 for testing. The performance comparison of the LAMSTAR network versus the BP network are also summarized in Table 13.B.1 above. As we observe, the LAMSTAR network is not only much faster to train (over 1000 times), but clearly gives better prediction accuracy (85% as compared to 78% for BP networks) with less sensitivity.

(c) Application to Diagnosis of Drug Abuse for Emergency Cases

In this application, the LAMSTAR network is used as a decision support system to identify the type of drug used by an unconscious patient who is brought to an emergency-room (data obtained from Maha Noujeime, University of Illinois at Chicago [Beirut et al., 1998, Noujeime, 1997]). A correct and very rapid identification of the drug type, will provide the emergency room physician with the immediate treatment required under critical conditions, whereas wrong or delayed identification may prove fatal and when no time can be lost, while the patient is unconscious and cannot help in identifying the drug. The LAMSTAR system can diagnose to distinguish between five groups of drugs: alcohol, cannabis (marijuana), opiates (heroin, morphine, etc.), hallucinogens (LSD), and CNS stimulants (cocaine) [Beirut et al., 1998]. In the drug abuse identification problem diagnosis can not be based on one or two symptoms since in most cases the symptoms overlap. The drug abuse identification is very complex problem since most of the drugs can cause opposite symptoms depending on additional factors like: regular/periodic use, high/low dose, time of intake [Beirut et al., 1998]. The diagnosis is based on a complex relation between 21 input variables arranged in 4 categories (subword vectors) representing drug abuse symptoms. Most of these variables are easily detectable in an emergency-room setting by simple evaluation (Table 2). The large number of variables makes it often difficult for a doctor to properly interrelate them under emergency room conditions for a correct diagnosis. An incorrect diagnosis, and a subsequent incorrect treatments may be lethal to a patient. For example, while cannabis and cocaine require different treatment, when analyzing only mental state of the patient, both cannabis and large doses of cocaine can result in the same mental state classified as mild panic and paranoia. Furthermore, often not all variables can be evaluated for a given patient. In emergency-room setting it is impossible to determine all 21 symptoms, and there is no time for urine test or other drug tests.

The LAMSTAR network was trained with 300 sets of simulated input data of the kind considered in actual emergency room situations [Kordylewski et al., 1999]. The testing of the network was performed with 300 data sets (patient cases), some of which have incomplete data (in emergency-room setting there is no time for urine or other drug tests). Because of the specific requirements of the drug abuse identification problem (abuse of cannabis should never be mistakenly identified as any other drug), the training of the system consisted of two phases. In the first phase, 200 training sets were used for unsupervised training, followed by the second phase where 100 training sets were used in on-line supervised training .

The LAMSTAR network successfully recognized 100% of cannabis cases, 97% of CNS stimulants, and hallucinogens (in all incorrect identification cases both drugs were mistaken with alcohol), 98% of alcohol abuse (2% incorrectly recognized as opiates), and 96% of opiates (4% incorrectly recognized as alcohol).

(d) Application to Assessing of Fetal Well-Being

This application [Scarpazza et al., 2002] is to determine neurological and cardiologic risk to a fetus prior to delivery. It concerns situations where, in the hours before delivery, the expectant mother is connected to standard monitors of fetal heart rate and of maternal uterine activity. Also available are maternal and other related clinical records. However, unexpected events that may endanger the fetus, while recorded, can reveal themselves over several seconds in one monitor and are not

Table 13.B.2. Symptoms divided into four categories for drug abuse diagnosis problem.

CATEGORY 1	CATEGORY 2	CATEGORY 3	CATEGORY 4
Respiration	Pulse	Euphoria	Physical Dependence
Temperature	Appetite	Conscious Level	Psychological Dependence
Cardiac Arrhythmia	Vision	Activity Status	Duration of Action
Reflexes	Hearing	Violent Behavior	Method of Administration
Saliva Secretion	Constipation	Convulsions	Urine Drug Screen

conclusive unless considered in the framework of data in anther monitor and of other clinical data. Furthermore, there is no expert physician available to constantly read any such data, even from a single monitor, during the several hours prior to delivery. This causes undue emergencies and possible neurological damage or death in approximately 2% of deliveries. In [Scarpazza et al., 2002] preliminary results are given where all data above are fed to a LAMSTAR neural network, in terms of 126 features, including 20 maternal history features, 9 maternal condition data at time of test (body temperature, number of contractions, dilation measurements, etc.) and 48 items from preprocessed but automatically accessed instruments data (including fetal heart rate, fetal movements, uterine activity and cross-correlations between the above).

This study on real data involved 37 cases used for training the LAMSTAR NN and 36 for actual testing. The 36 test cases involved 18 positives and 18 negatives. Only one of the positives (namely, indicating fetal distress) was missed by the NN, to yield a 94.44% sensitivity (miss-rate of 5.56%). There were 7 false alarms as is explained by the small set of training cases. However, in a matter of fetal endangerment, one obviously must bias the NN to minimize misses at the cost of higher rate of false alarms. Computation time is such that decisions can be almost real time if the NN and the preprocessors involved are directly connected to the instrumentation considered.

Several other applications to this problem were reported in the literature, using other neural networks [Scarpazza et al., 2002]. Of these, results were obtained in [Rosen et al., 1997] where the miss percentage (for the best of several NN's discussed in that study) was reported as 26.4% despite using 3 times as many cases for NN-training. Studies based on Back-Propagation yielded accuracy of 86.3% for 29 cases over 10.000 iterations and a miss rate of 20% [Maeda et al., 1998], and 11.1% miss-rate using 631 training cases on a test set of 319 cases with 15,000 iterations [Kol et al. 1995].

Problems

Chapter 1:

Problem 1.1:

Explain the major difference between a conventional (serial) computer and a neural network.

Problem 2.2:

Explain the points of difference between a mathematical simulation of a biological neural system (central nervous system) and an artificial neural network.

Chapter 2:

Problem 2.1:

Reconstruct a fundamental input/nonlinear-operator/output structure of a biological neuronal network, stating the role of each element of the biological neuron in that structure.

Chapter 3:

Problem 3.1:

Explain the difference between the LMS algorithm of Sec. 3.4.1 and the gradient algorithm of Sec. 3.4.2. What are their relative merits?

Problem 3.2:

Intuitively, explain the role of μ in the gradient algorithm, noting its form as in Eq. (3.25).

Chapter 4:

Problem 4.1:

Compute the weights $\mathbf{w}_k$ of a single perceptron to minimize the cost J where $J = E[(d_k - z_k)^2]$; E denoting expectation, d_k being the desired output and z_n denoting the net (summation) output, and where $z_k = \mathbf{w}_k^T \mathbf{x}_n$; $\mathbf{x}_k$ being the input vector.

Problem 4.2:

Explain why a single-layer perceptron cannot solve the XOR problem. Use an X1 vs. X2 plot to show that a straight line cannot separate the XNOR states.

Problem 4.3:

Explain why a 2-layer perceptron can solve the XOR problem.

Chapter 5:

Problem 5.1:

Design a 2-layer Madaline neural network to recognize digits from a set of 3 cursive digits in a 5-by-5 grid.

Chapter 6:

Problem 6.1:

Design a back propagation (BP) network to solve the XOR problem.

Problem 6.2:

Design a BP network to solve the XNOR problem.

Problem 6.3:

Design a BP network to recognize handwritten cursive digits from a set of 3 digits written in a 10-by-10 grid.

Problem 6.4:

Design a BP network to recognize cursive digits as in Prob. 6.3, but with additive single-error-bit noise.

Problem 6.5:

Design a BP network to perform a continuous wavelet transform $W(\alpha, \tau)$ where:

$$W(\alpha, \tau) = \frac{1}{\sqrt{\alpha}} \int f(t) g\left(\frac{t - \tau}{\alpha}\right) dt$$

$f(t)$ being a time-domain signal, α denoting scaling, and τ denoting translation.

Chapter 7:

Problem 7.1:

Design a Hopfield network to solve a 6-city and an 8-city Travelling-Salesman problem. Give solutions for 30 iterations. Determine distances as for 6 or 8 cities arbitrarily chosen from a US road-map, to serve as your distance matrix.

Problem 7.2:

Design a Hopfield network to recognize handwritten cursive characters from a set of 5 characters in a 6-by-6 grid.

Problem 7.3:

Repeat Prod. 7.2 for added single-error-bit noise.

Problem 7.4:

Repeat Problem 7.2 for added two-bit noise.

Problem 7.5:

Explain why w_{ii} must be 0 for stability of the Hopfield Network.

Chapter 8:

Problem 8.1:

Design a Counter-Propagation (CP) network to recognize cursive handwritten digits from a set of 6 digits written in an 8-by-8 grid.

Problem 8.2:

Repeat Prob. 8.1 for added single-error-bit noise.

Problem 8.3:

Explain how the CP network can solve the XOR problem.

Chapter 9:

Problem 9.1:

Design an ART-I network to recognize cursive handwritten characters from a set of 6 characters written in an 8-by-8 grid.

Problem 9.2:

Repeat Prob. 9.1 for added single-error-bit noise.

Chapter 10:

Problem 10.1:

Explain how competition is accomplished in Cognitron networks and what its purpose is.

Problem 10.2:

Explain the difference between the S and the C layers in a neocognitron network and comment on their respective roles.

Chapter 11:

Problem 11.1:

Design a sochastic BP network to recognize cursive characters from a set of 6 characters written in an 8-by-8 grid.

Problem 11.2:

Repeat Prob. 11.1 for characters with additive single-error-bit noise.

Problem 11.3:

Design a stochastic BP network to identify the autoregressive (AR) parameters of an AR model of a discrete-time signal x_k given by:

$$x_k = a_1 x_{k-1} + a_2 x_{k-2} + w_k; \quad k = 1, 2, 3, \ldots$$

where $a_1; a_2$ are the AR parameters to be identified.
Generate the signal by using a model (unknown to the neural network), as follows:

$$x_k = 0.5x_{k-1} - 0.2x_{k-2} + w_k$$

where w_k is Gaussian white noise.

Chapter 12:

Problem 12.1:

Design a recurrent BP network to solve the AR-identification problem as in Prob. 11.3.

Problem 12.2:

Repeat Prob. 12.1 when employing simulated annealing in the recurrent BP network.

Chapter 13:

Problem 13.1:

In applying the LAMSTAR network to a diagnostic situation as in Sec. 13.A, when a piece of diagnostic information (a subword) is missing, what will the display for that subword show?

Problem 13.2:

In applying the LAMSTAR network to a diagnostic situation as in Sec. 13.A, when a certain diagnostic information (a subword) is missing, can the LAMSTAR network still yield a diagnostic decision-output? If so, then how?

Problem 13.3:

(a) Given examples of several (4 to 6) input words and of their subwords and briefly explain the LAMSTAR process of troubleshooting for a simple fault-diagnosis problem involved in detecting why your car does not start in the morning. You should set up simulated scenarios based on your personal experiences.
(b) What should a typical output word be? Explain how the CP network can solve the XNOR problem.

References

Allman, J., Miezen, F., and McGuiness, E. [1985] "Stimulus specific responses from beyond the classical receptive field", *Annual Review of Neuroscience* **8**, 147–169.

Barlow, H. [1972] "Single units and sensation: A neuron doctrine for perceptual psychology", *Perception* **1**, 371–392.

Barlow, H. [1994] "What is the computational goal of the neocortex?", *Large-Scale Neuronal Theories of the Brain*, MIT Press, Chap. 1.

Bear, M. F., Cooper, L. N. and Ebner, F. E. [1989] "A physiological basis for a theory of synapse modification", *Science*, **237**, pp. 42–47.

Beauchamp, K. G. [1984] *Applications of Walsh and Related Functions*, Academic Press, London.

Beauchamp, K. G. [1987] "Sequency and Series, in *Encyclopedia of Science and Technology*, ed. Meyers, R. A., Academic Press, Orlando, FL, pp. 534–544.

Bellman, R. [1961], *Dynamic Programming*, Princeton Univ. Press, Princeton, N.J.

Bierut, L. J. *et al.* [1998], "Familiar transmission of substance dependence: alcohol, marijuana, cocaine, and habitual smoking", *Arch. Gen. Psychiatry*, **55**(11), pp. 982–988.

Carino, C., Lambert, B., West, P. M., Yu, C. [2005], Mining officially unrecognized side effects of drugs by combining web search and machine learning — *Proceedings of the 14th ACM International Conference on Information and Knowledge Management.*

Carlson, A. B. [1986] *Communications Systems*, McGraw Hill, New York.

Carpenter, G. A. and Grossberg, S. [1987a] "A massively parallel architecture for a self-organizing neural pattern recognition machine", *Computer Vision, Graphics, and Image Processing*, **37**, 54–115.

Carpenter, G. A. and Grossberg, S. [1987b] "ART-2: Self-organizing of stable category recognition codes for analog input patterns", *Applied Optics* **26**, 4919–4930.

Cohen, M. and Grossberg, S. [1983] "Absolute stability of global pattern formation and parallel memory storage by competitive neural networks", *IEEE Trans. Sys., Man and Cybernet.* **SMC-13**, 815–826.

Cooper, L. N. [1973] "A possible organization of animal memory and learning", *Proc.*

Nobel Symp. on Collective Properties of Physical Systems, ed. Lundquist, B. and Lundquist, S., Academic Press, New York, pp. 252–264.

Crane, E. B. [1965] *Artificial Intelligence Techniques*, Spartan Press, Washington, DC.

Ewing, A. C. [1938] *A Short Commentary of Kant's Critique of Pure Reason*, Univ. of Chicago Press.

Fausett, L. [1993] *Fundamentals of Neural Networks, Architecture, Algorithms and Applications*, Prentice Hall, Englewood Cliffs, N.J.

Freeman, J. A. and Sakpura, D. M. [1991] *Neural Nentworks, Algorithms, Applications and Programming Techniques*, Addison Wesley, Reading, MA.

Fukushima, K. [1975] "Cognitron, a self-organizing multilayered neural network", *Biological Cybernetics*, **20**, pp. 121–175.

Fukushima, K., Miake, S. and Ito, T. [1983] "Neocognitron: a neural network model for a mechanism of visual pattern recognition", *IEEE Trans. on Systems, Man and Cybernetics*, **SMC-13**, 826–834.

Ganong, W. F. [1973] *Review of Medical Physiology*, Lange Medical Publications, Los Altos, CA.

Gee, A. H. and Prager, R. W. [1995], Limitations of neural networks for solving traveling salesman problems, *IEEE Trans. Neural Networks*, vol. 6, pp. 280–282.

Geman, S. and Geman, D. [1984] "Stochastic relaxation, Gibbs distributions, and the Bayesian restoration of images", *IEEE Trans. on Pattern Anal. and Machine Intelligence* **PAM1-6**, 721–741.

Gilstrap, L. O., Lee, R. J. and Pedelty, M. J. [1962] "Learning automata and artificial intelligence", in *Human Factors in Technology*, ed. Bennett, E. W. Degan, J. and Spiegel, J., McGraw-Hill, New York, NY, pp. 463–481.

Girado, J. I., Sandin, D. J., DeFanti T. A., Wolf, L. K. [2003], Real-time Camera-based Face Detection using a Modified LAMSTAR Neural Network System — *Proceedings of IS&T/SPIE's 15th Annual Symposium Electron. Imaging.*

Graupe, D. [1997] *Principles of Artificial Neural Networks*, World Scientific Publishing Co., Singapore and River Edge, N.J., (especially, chapter 13 thereof).

Graupe, D. [1989] *Time Series Analysis, Identification and Adaptive Filtering*, second edition, Krieger Publishing Co., Malabar, FL.

Graupe, D. and Kordylewski, H. [1995] "Artificial neural network control of FES in paraplegics for patient responsive ambulation", *IEEE Trans. on Biomed. Eng.*, **42**, pp. 699–707.

Graupe, D. and Kordylewski, H. [1996a] "Network based on SOM modules combined with statistical decision tools", *Proc. 29th Midwest Symp. on Circuits and Systems*, Ames, IO.

Graupe, D. and Kordylewski, H. [1996b] "A large memory storage and retrieval neural network for browsing and medical diagnosis applications", in "Intelligent Engineering Systems Through Artificial Neural Networks", Vol. 6, Editors: Dagli, C. H., Akay, M., Chen, C. L. P., Fernandez, B. and Ghosh, J., *Proc. Sixth ANNIE Conf.*, St. Louis, MO, ASME Press, NY, pp. 711–716.

Graupe, D. and Kordylewski, H. [1998] A Large Memory Storage and Retrieval Neural Network for Adaptive Retrieval and Diagnosis, *Internat. J. Software Eng. and Knowledge Eng.*, Vol. 8, No. 1, pp. 115–138.

Graupe, D. and Kordilewski, H., [2001], A Novel Large-Memory Neural Network as an Aid in Medical Diagnosis, *IEEE Trans. on Information Technology in Biomedicine*, Vol. 5, No. 3, pp. 202–209, Sept. 2001.

Graupe, D., Lynn, J. W. [1970] "Some aspects regarding mechanistic modelling of recognition and memory", *Cybernetica*, **3**, 119–141.

Grossberg, S. [1969] "Some networks that can learn, remember and reproduce any number of complicated space-time patterns", *J. Math. and Mechanics*, **19**, 53–91.

Grossberg, S. [1974] "Classical and instrumental learning by neural networks", *Progress in Theoret. Biol.*, **3**, 51–141, Academic Press, New York.

Grossberg, S. [1982] "Learning by neural networks", in *Studies in Mind and Brain*, ed. Grossberg, S., D. Reidel Publishing Co., Boston, MA., pp. 65–156.

Grossberg, S. [1987] "Competitive learning: From interactive activation to adaptive resonance", *Cognitive Science* **11**, 23–63.

Guyton, A. C. [1971] *Textbook of Medical Physiology*, 14th edition, W. B. Saunders Publ. Co., Philadelphia.

Hammer, A., Lynn, J. W., and Graupe, D. [1972] "Investigation of a learning control system with interpolation", *IEEE Trans. on System Man, and Cybernetics* **2**, 388–395.

Hammerstrom, D. [1990] "A VLSI architecture for high-performance low-cost on-chip learning", *Proc. Int. Joint Conf. on Neural Networks*, vol. 2, San Diego, CA, pp. 537–544.

Hamming, R. W. [1950] *Error Detecting and Error Correcting Codes*, Bell Sys. Tech. J. **29**, 147–160.

Happel, B. L. M., Murre, J. M. J. [1994] "Design and evolution of modular neural network architectures", *Neural Networks* **7**, 7, 985–1004.

Harris, C. S. [1980] "Insight or out of sight?" Two examples of perceptual plasticity in the human", *Visual Coding and Adaptability*, 95–149.

Haykin, S., [1994] *Neural Networks, A Comprehensive Foundation*, Macmillan Publ. Co., Englewood Cliffs, NJ.

Hebb, D. [1949] *The Organization of Behavior*, John Wiley, New York.

Hecht-Nielsen, R. [1987] "Counter propagation networks", *Applied Optics* **26**, 4979–4984.

Hecht-Nielsen, R. [1990] *Neurocomputing*, Addison-Wesley, Reading, MA.

Hertz, J., Krogh, A. and Palmer, R. G. [1991] *Introduction to the Theory of Neural Computation*, Addison-Wesley, Reading, MA.

Hinton, G. E. [1986] "Learning distributed representations of concepts", *Proc. Eighth Conf. of Cognitive Science Society* vol. 1, Amherst, MA.

Hopfield, J. J. [1982] "Neural networks and physical systems with emergent collective computational abilities", *Proceedings of the National Academy of Sciences* **79**, 2554–2558.

Hopfield, J. J. and Tank, D. W. [1985], Neural computation of decisions in optimization problems, *Biol. Cybern.*, vol. 52, pp. 141–152.

Hubel, D. H. and Wiesel, T. N. [1979] "Brain mechanisms of vision", *Scientific American* **241**, 150–162.

Jabri, M. A., Coggins, R. J. and Flower, B. G. [1996] *Adaptive Analog VLSI Neural Systems*, Chapman and Hall, London.

Kaski, S., Kohonen, T. [1994] "Winner-take-all networks for physiological models of competitive learning", *Neural Networks* **7** (7) 973–984.

Katz, B. [1966] *Nerve, Muscle and Synapse*, McGraw-Hill, New York.

Kohonen, T. [1977] *Associated Memory: A System-Theoretical Approach*, Springer Verlag, Berlin.

Kohonen, T. [1984] *Self-Organization and Associative Memory*, Springer Verlag, Berlin.

Kohonen, T. [1988] "The neural phonetic typewriter", *Computer* **21**(3).

Kohonen, T. [1988] *Self Organizing and Associative Memory*, second edition, Springer Verlag, New York.

Kol, S., Thaler, I., Paz, N. and Shmueli, O. [1995], Interpretation of Nonstress Tests by an Artificial NN, *Amer. J. Obstetrics & Gynecol.*, **172**(5), 1372–1379.

Kordylewski, H. and Graupe, D. [1997] Applications of the LAMSTAR Neural Network to Medical and Engineering Diagnosis/Fault Detection, *Proc. 7th ANNIE Conf.*, St. Louis, MO.

Kordylewski, H. [1998], A Large Memory Storage and Retrieval Neural Network for Medical and Industrial Diagnosis, Ph.D. Thesis, EECS Dept., Univ. of Illinois, Chicago.

Kordylewski, H., Graupe, D. and Liu, K. [1999] Medical Diagnosis Applications of the LAMSTAR Neural Network, *Proc. of Biol. Signal Interpretation Conf.* (*BSI-99*), Chicago, IL.

Kosko, B. [1987] "Adaptive bidirectional associative memories", *Applied Optics* **26**, 4947–4960.

Lee, R. J. [1959] "Generalization of learning in a machine", *Proc. 14th ACM National Meeting*, September.

Levitan, L. B., Kaczmarek, L. K. [1997] The Neuron, Oxford Univ. Press, 2nd Ed.

Livingstone, M., and Hubel, D. H. [1988] "Segregation of form, color, movement, and depth: Anatomy, physiology, and perception", *Science* **240**, 740–749.

Longuett-Higgins, H. C. [1968] "Holographic model of temporal recall", *Nature* **217**, 104.

Lyapunov, A. M. [1907] "Probléme général de la stabilité du mouvement", *Ann. Fac. Sci. Toulouse* **9**, 203–474; English edition: *Stability of Motion*, Academic Press, New York, 1957.

Martin, K. A. C. [1988] "From single cells to simple circuits in the cerebral cortex", *Quart. J. of Experimental Physiology* **73**, 637–702.

Maeda, K., Utsu, M., Makio, A., Serizawa, M., Noguchi, Y., Hamada, T., Mariko, K. and Matsumo, F. [1998], Neural Network Computer Analysis of Fetal Heart Rate, *J. Maternal-Fetal Investigation* **8**, 163–171.

McClelland, J. L. [1988] "Putting knowledge in its place: a scheme for programming parallel processing structures on the fly", in *Connectionist Models and Their Implication*, Chap. 3, Ablex Publishing Corporation.

McCulluch, W. S. and Pitts, W. [1943] "A logical calculus of the ideas imminent in nervous activity", *Bulletin Mathematical Biophysics*, 5, 115–133.

Metropolis, N., Rosenbluth, A. W., Rosenbluth, M. N., Teller, A. H. and Teller, E. [1953] "Equations of state calculations by fast computing machines", *J. Chemistry and Physics* **21**, 1087–1091.

Minsky, M. L. [1980] "K-lines: A theory of memory", *Cognitive Science* **4**, 117–133.

Minsky, M. L. [1987] "*The society of mind*", Simon and Schuster, New York.

Minsky, M. L. [1991] "Logical versus analogical or symbolic versus neat versus scruffy", *AI Mag.*

Minsky, M. and Papert, S. [1969] *Perceptrons*, MIT Press, Cambridge, MA.

Morrison, D. F. [1996], *Multivariate Statistical Methods*, McGraw-Hill, p. 222.

Mumford, D. [1994] "Neural architectures for pattern-theoretic problems", *Large-Scale Neuronal Theories of the Brain*, Chap. 7, MIT Press.

Niederberger, C. S. *et al.* [1996], A neural computational model of stone recurrence after ESWL, *Internat. Conf. on Eng. Appl. of Neural Networks* (*EANN '96*), pp. 423–426.

Noujeime, M. [1997], *Primary Diagnosis of Drug Abuse for Emergency Case*, Project Report, EECS Dept., Univ. of Illinois, Chicago.

Nii, H. P. [1986] "Blackboard systems: Blackboard application systems", *AI Mag.* **7**, 82–106.

Parker, D. B. [1982] *Learning Logic*, Invention Report 5-81-64, File 1, Office of Technology Licensing, Stanford University, Stanford, CA.

Patel, T. S. [2000], *LAMSTAR NN for Real Time Speech Recognition to Control Functional Electrical Stimulation for Ambulation by Paraplegics*, MS Project Report, EECS Dept., Univ. of Illinois, Chicago.

Pineda, F. J. [1988] "Generalization of backpropagation to recurrent and higher order neural networks", pp. 602–611, in *Neural Information Processing Systems*, ed. Anderson, D. Z. Amer. Inst. of Physics, New York.

Poggio, T., Gamble, E. B., and Little, J. J. [1988] "Parallel integration of vision modules", *Science* **242**, 436–440.

Riedmiller, M. and Braun, H. [1993], "A direct adaptive method for faster backpropagation learning: The RPROP algorithm", *Proc. IEEE Conf. Neur. Networks*, 586–591, San Francisco.

Rosen, B. E., Bylander, T. and Schifrin, B. [1997], Automated diagnosis of fetal outcome from cardio-tocograms, *Intelligent Eng. Systems Through Artificial Neural Networks*, NY, ASME Press, **7**, 683–689.

Rosenblatt, F. [1958] "The perceptron, a probabilistic model for information storage and organization in the brain", *Psychol. Rev.* **65**, 386–408.

Rosenblatt, F. [1961] *Principles of Neurodynamics, Perceptrons and the Theory of Brain Mechanisms*, Spartan Press, Washington, DC.

Rumelhart, D. E., Hinton, G. E. and Williams, R. J. [1986] "Learning internal representations by error propagation", pp. 318–362 in *Parallel Distributed Processing: Explorations in the Microstructures of Cognition*, eds. Rumelhart, D. E. and McClelland, J. L. MIT Press, Cambridge, MA.

Rumelhart, D. E. and McClelland, J. L. [1986] "An interactive activation model of the effect of context in language learning", *Psychological Review* **89**, 60–94.

Sage, A. P. and White, C. C., III [1977] *Optimum Systems Control*, second edition, Prentice Hall, Englewood Cliffs, NJ.

Scarpazza, D. P., Graupe, M. H., Graupe, D. and Hubel, C. J. [2002], Assessment of Fetal Well-Being Via A Novel Neural Network, *Proc. IASTED International Conf. On Signal Processing, Pattern Recognition and Application*, Heraklion, Greece, pp. 119–124.

Sejnowski, T. J. [1986] "Open questions about computation in cerebral cortex", *Parallel Distributed Processing*, **2**, 167–190.

Sejnowski, T. J. and Rosenberg, C. R. [1987] "Parallel networks that learn to pronounce English text", *Complex Systems* **1**, 145–168.

Singer, W. [1993] "Synchronization of cortical activity and its putative role in information processing and learning", *Ann. Rev. Physiol.* **55**, 349–374.

Smith, K., Palaniswami, M. and Krishnamoorthy, M. [1998], *Neural Techniques for Combinatorial Optimization with Applications*, vol. 9, no. 6, pp. 1301–1318.

Szu, H. [1986] "Fast simulated annealing", in *Neural Networks for Computing*, ed. Denker, J. S. Amer. Inst. of Physics, New York.

Thompson, R. F. [1986] "The neurobiology of learning and memory", *Science*, **233**, 941–947.

Todorovic, V. [1998], *Load Balancing in Distributed Computing*, Project Report, EECS Dept., Univ. of Illinois, Chicago.

Ullman, S. [1994] "Sequence seeking and counterstreams: A model for bidirectional information flow in the cortex", *Large-Scale Neuronal Theories of the Brain*, Chap. 12 MIT Press.

Waltz, D. and Feldman, J. [1988] *Connectionist Models and Their Implication*, Ablex Publishing Corporation.

Wasserman, P. D. [1989] *Neural Computing; Theory and Practice*, Van Nostrand Reinhold, New York.

Werbos, P. J. [1974] "Beyond recognition; new tools for prediction and analysis in the behavioral sciences", Ph.D. Thesis, Harvard Univ., Cambridge, MA.

Widrow, B. and Hoff, M. E. [1960] "Adaptive switching circuits", *Proc. IRE WESCON Conf.*, New York, pp. 96–104.

Widrow, B. and Winter, R. [1988] "Neural nets for adaptive filtering and adaptive pattern recognition", *Computer* **21**, pp. 25-39.

Wilks, S. [1938] "The large sample distribution of the likelihood ration for testing composite hypothesis", *Ann. Math. Stat.*, Vol. 9, pp. 2–60.

Wilson, G. V. and Pawley, G. S. [1998], On the stability of the TSP algorithm of Hopfield and Tank, *Biol. Cybern.*, vol. 58, pp. 63–70.

Windner, R. O. [1960] "Single storage logic", *Proc. AIEE*, Fall Meeting, 1960.

Author Index

Subject Index